D1267885

The Science of Monsters

The Origins of the Creatures We Love to Fear

Previously published as *Medusa's Gaze and Vampire's Bite*

Matt Kaplan

Scribner
New York London Toronto Sydney New Delhi

To John Leopold
for bringing monsters into my life

To Thalia and my parents
for putting up with them around the house

Scribner
A Division of Simon & Schuster, Inc.
1230 Avenue of the Americas
New York, NY 10020

Copyright © 2012 by Matt Kaplan
"Afterword" © 2013 by Matt Kaplan

All rights reserved, including the right to reproduce this book or portions thereof in any form whatsoever. For information address Scribner Subsidiary Rights Department, 1230 Avenue of the Americas, New York, NY 10020.

First Scribner trade paperback edition October 2013
Previously published as *Medusa's Gaze and Vampire's Bite* by Scribner.

SCRIBNER and design are registered trademarks of The Gale Group, Inc., used under license by Simon & Schuster, Inc., the publisher of this work.

For information about special discounts for bulk purchases, please contact Simon & Schuster Special Sales at 1-866-506-1949 or business@simonandschuster.com.

The Simon & Schuster Speakers Bureau can bring authors to your live event. For more information or to book an event contact the Simon & Schuster Speakers Bureau at 1-866-248-3049 or visit our website at www.simonspeakers.com.

Designed by Carla Jayne Jones

Manufactured in the United States of America

10 9 8 7 6 5 4 3 2

Library of Congress Control Number: 2012016553

ISBN 978-1-4516-6798-1
ISBN 978-1-4516-6799-8 (pbk)
ISBN 978-1-4516-6800-1(ebook)

The lines from the book of Job on page 85 are from the Holy Bible, New International Version®, NIV®, copyright © 1973, 1978, 1984, 2011 by Biblica, Inc.™ Used by permission. All rights reserved worldwide.

R0453151317

Contents

Acknowledgments

As a science journalist, your life has a steady rhythm to it. Wednesday and Thursday, read new papers on everything from the chemistry of wine to the behavior of rare parasites. Friday, write up summaries of the most interesting papers that you've found and send these off to editorial. Sunday evening, get your marching orders. Monday, write articles on the papers that editorial liked and communicate with the scientists behind these papers to make sure you aren't getting anything wrong. Tuesday, proofread alterations made by editorial and approve publication. Wednesday, do it all over again.

You get exposed to a lot of fascinating and cutting-edge science over the years in this line of work (great for cocktail parties*) but never develop the sort of expertise that a professional researcher might. For this reason I am deeply indebted to the small legion of anthropologists, archaeologists, astrobiologists, chemists, computer scientists, ecologists, engineers, epidemiologists, geologists, mariners, medical doctors, microbiologists, oceanographers, parasitologists, paleontologists, psychologists, and zoologists who stepped in and double-checked my work. Anja Scheffers, Arne Öhman, Blaire Van Valkenburgh, Charles Stewart, Chris Mark, Claudia Montiel-

*Or so I hear. None of my friends throw cocktail parties. Maybe I'll throw one when this book gets published just to see if all the stuff in my head is of any use.

Equihua, Curtis Marean, Daniel Bartels, David Hankin, David Hughes, Eli Finkel, Elizabeth Lonsdorf, Elizabeth Hadly, Harry Greene, Howard Spero, John Hutchinson, Judith Morales, Lee Kagan, Louis-Philippe Morency, Márta Korbonits, Michael Benton, Michael Russell, Nathanael Fast, Netsebrak Zewde, Norman Macleod, Paul Eastwick, Paul Rozin, Randy Hill, Ray Kay, Richard Cowen, Robert Ness, and Tony Barnosky, thank you so much for all of your time and effort. To those of you whom I have collaborated with over the years on articles for *The Economist, Nature, New Scientist,* and *National Geographic,* it has been a pleasure.

Of course, *Medusa's Gaze and Vampire's Bite* explores much more than science. Ancient history and myths are critical to this book and, while I have a soft spot in my heart for these subjects, I studied paleontology at university, not classics. I am thus very grateful to Adrienne Mayor and John Leopold for so meticulously going through the manuscript and offering such valuable advice and support on all things Greek. *Epharisto!*

Many thanks must also go to Alex Caley, Anna Hankin, Anne Garry, Anthony Hartley, Charlie Hsu, David Attenborough, Evan Mooney, Hsien-I Chien, Judith Seidel, Kate Coe, Matthew Johnson, Nathaniel Bottrell, Rafal Marszalek, Sharon Dewar, and Wilfred Yung for having the courage to tell me when the road through my text got bumpy, and to my agent, Erin Malone, for her helpful feedback throughout this process.

Finally, I owe my editor, Anna DeVries, a great deal of gratitude for convincing me to write this book in the first place.

Introduction

In the darkness it came. There was no way out. Cornered and helpless, all who found themselves in this dreaded place knew their fate. Relentlessly, the half-human, half-bull fiend found its quarry and tore them to pieces. Scrambling and searching for an exit was pointless. Even if any did miraculously find a way out, the natives of the island were against them. They would simply throw escapees back into the blackness of the depths. Whether their deaths were quick or drawn out remains a mystery, for in the labyrinth of legend on the island of Crete, none of those forced in were ever heard from again.

To those who feel the Minotaur is too ancient to be relevant anymore, consider the fate of the mining vessel *Nostromo* in Ridley Scott's *Alien.* Stalked relentlessly by a carnivorous beast, one by one the crew are ripped apart and consumed. It is impossible to find on the shadowy ship, and it makes its attacks almost entirely unseen. Computers, bullets, and flamethrowers are useless. The alien sprays acidic blood when cut that disintegrates armor, burns flesh, and melts bone. There is no escape and no rescue on the way, because "in space, no one can hear you scream."

Defined as horrible to behold and a threat to all who cross them, monsters are creatures we run from and beasts we warn our children about. Yet something about them is enticing, mesmerizing, and addictive. Terrible as they might be, we cannot help looking ever closer, parting the fingers that are covering our eyes. There is no getting around it: Something deep inside monsters fascinates us.

What it is about monsters that is so alluring is hard to say. Seeing them makes the heart pump faster, hairs stand on end, and sweat pour down our faces. All of these are signs of stress and are often experienced over and over again through nightmares. Even so, children clamor for ghost stories around the campfire and adults line up in droves to see films featuring vampires and werewolves. They terrify, yet we cannot get enough of them.

And this is nothing new. The Minotaur, Sphinx, and Medusa were created long ago, and based upon their representation in myths, poems, art, and plays, it seems that they drew attention from ancient audiences that was very much like the attention drawn by modern monsters. This hints that monsters have been with us for quite a long time and raises a perplexing question: Why have monster stories, which have the effect of scaring people, persisted so relentlessly throughout the ages?

The masochism tango

As bizarre as it sounds, one answer to this question lies with research on why people like spicy foods. Dishes from Mexico and India are tongue-searingly hot. They make your eyes burn and can soak you with sweat. A lot of people avoid them, but many love them precisely because they are so fiery.

It defies logic that food responsible for such a seemingly painful experience should be so popular, but recent work is beginning to provide an explanation. Fascinated and befuddled by the common human desire to eat mouth-burning foods, psychologist Paul Rozin and a team of colleagues at the University of Pennsylvania wondered whether it was the negative experience of being burned that spicy-food lovers liked or if it was their body's physiological reactions to these foods that they were enjoying.

The team asked 135 female and 108 male university students

to rate on a scale of 0 to 100 how much they liked different things, with 0 indicating "not liking at all" and 100 indicating "considerable liking." When the students were asked, among other things, how much they liked spicy foods, the average score was 55.5, which runs roughly along with the perception that around half of the population enjoys this sort of cuisine. However, far more interesting was that when the participants were asked to rate how much they enjoyed mouth burns, sweating, and tearing eyes, those who rated their love of spicy food at over 50 also tended to rate these typically unpleasant experiences more highly. This suggested they were actually enjoying their body's own negative response to the food.

The reason for this masochism is not known, but Rozin, along with many others in his field, have a theory that there is pleasure for the mind in watching the body react negatively while knowing perfectly well that nothing bad is actually going to happen. The enjoyment, they suggest, comes from a sense of mental mastery over the body that is responding in a knee-jerk reaction.

Rozin's study did not limit itself to an exploration of spicy cuisine. It also asked participants to rate how much they enjoyed thrill rides, frightening movies, gory movies, and even a pounding heart. Again, a connection was found. Those who enjoyed getting the crap scared out of them in movies also tended to like gore, thrill rides, and a pounding heart. Mental mastery might be behind this too.

Just as the brain is able to identify that screaming taste buds are screaming about nothing serious, the brain is capable of realizing that a frightening story is not real. Researchers propose that in this realization there is a sense of mastery of mind over body that is, in itself, enjoyable.

So where does all of this place monsters? Well, they are by their very nature frightening. For some, like Medusa, the fear factor is in their physical nature; they literally are scary to look upon (just think about the number of people who are petrified by snakes). For others, like the vampire, it is their activities, like the sucking of blood from living victims, that engender feelings of fear. And then there are those, like the Minotaur and the alien, that elicit a feeling of dread by forcing us back into the ancient position of being prey as our ancestors once were. For these reasons it is possible that simply thinking about monsters reassures us we are above other animals in our control of fate much as chili peppers remind us that we, and not our taste buds, are in control of what we eat. True, none of the psychological research has specifically looked at monster movies or stories, but the connection seems logical. Even so, there is probably more that draws people to be fascinated by monsters than just pleasure. Monsters likely also serve a practical purpose.

Playing in the sandbox

At their most basic level, monsters represent fears held by society, fears associated with dangers perceived in the surrounding world. These fears have a powerful evolutionary history by encouraging people to flee instead of fighting suicidal battles. When ancient hunters encountered a saber-toothed tiger by accident, they ran. When the human ancestor *Homo erectus* caught angry cave bears by surprise, it ran. When chimpanzees and bonobos, the nearest genetic relatives to modern humans, encounter large predators in the wild, they run. While Hollywood heroes have made running away distinctly unpopular on the silver screen, every single actor who has ever portrayed a hero who stood his or her ground against some abominable terror comes from a long genetic lineage of cowards who fled in the face of danger. That is why they are here to act today. If their ancestors had fought against monsters far more powerful than themselves, as Hollywood heroes do all the time, their lineage would have been destroyed by predators long ago. Fear, in short, keeps people alive. But fear can also go too far.

Recent work in animal behavior has revealed something fascinating: There are personality types in animals. Among fish in a single species, there are adventurous individuals, ready and willing to take risks, and there are more cautious and timid individuals, fearful of doing anything that could put them in danger. Similar variations in personality are starting to be found in birds and mammals too. A recent study led by Kathryn Arnold at the University of York revealed that when greenfinches were presented with brightly colored objects in their food, there was considerable variation in how long it took each bird to eat. When intriguing objects were attached to the birds' perches, a similar variation was found. Some birds quickly flew to explore the new toy while others stayed away.

Being courageous or curious undoubtedly presents serious dangers. Ongoing studies indicate that fish with more daring personalities are more likely to nibble on bait on the end of a hook and risk-taking rodents more commonly end up in traps set by researchers.* Yet having a personality that predisposes an animal to take risks can yield rewards. Courage can lead an animal to investigate previously unexplored locations where food is present, or it can lead to the discovery of well-hidden nesting areas that have yet to be found by any other members of the species. Such discoveries can lead to better health and better breeding opportunities for the courageous animal that allow for its courageous genes to be passed along more readily to the next generation.

Whether some humans are genetically predisposed to be more adventurous than others remains to be determined, but there clearly are some people who ultimately are more willing to take risks. Make no mistake, the instinct to flee from danger is still deeply rooted in every person's brain, but some of us are more willing than others to go to places associated with danger. Just as with daring animals that find resources by taking risks, it is logical to assume that more adventurous humans have historically made the same sorts of gains. For this reason, monsters may be serving a valuable purpose in society. By representing key fears and allowing these to be discussed and explored in a safe

*As it happens, this has really screwed up lots of biological research. We have spent decades "thinking" we could get a reasonable sense of what animals are like by setting traps in the wild and then studying the animals that get caught. But if the animals that get caught are only the most daring individuals (or the most foolish) in a population, they are hardly giving us a reasonable sense of how a species behaves!

environment, monsters might be making it feasible for these fears to be more effectively prepared for and ultimately faced, so the benefits of being a courageous individual can be more readily reaped. Like lion cubs play-fighting in the safety of their den, monsters may be allowing threats to be toyed with in the safe sandbox of the imagination.

So if monsters are present in society for both pleasure and mental practice for future frightening interactions, what happens when our fears are overcome? What then?

To a certain extent, danger should function as the life essence of monsters. Once a perceived danger is dispelled, this essence is destroyed and the beast becomes extinct. It may continue to live on in fiction as a fossil of its former self or as a mere creature of interest, but not as a monster with all of the terror that comes with such status.

Fears have changed a lot since the dawn of humanity, and with these changes have come alterations in the pantheon of monsters that lurk in our world. The Minotaur is no longer with us, but aliens are. In a sense, monsters, while strictly the stuff of fantasy, experience evolution at a rate that is in stride with the pace of human understanding of the surrounding world. Science, the empirical testing and exploration of the world, which is about as seemingly unrelated to monsters as can be, is both responsible for their birth by discovering new environments where they might be living and the cause of their destruction through the ultimate revelation that they cannot possibly be real.

That many monsters have risen and fallen throughout the ages is clear. What is less clear is which specific fears these monsters stood for and how long these fears actually lasted. An exploration of fear's mask, the mask of the monster, seems an excellent way to find out.*

---

*A gentle disclaimer as we begin our little tour of all things monstrous: There is no way to know with certainty what was actually in our ancestors' minds when they invented the various monsters that have come to haunt our world. We have to guess. Guesses come in many forms. There are those of the wild variety and there are the educated ones guided by shreds of evidence. You will find the latter in this book.

1

Giant Animals—Nemean Lion, Calydonian Boar, Rukh, King Kong

"Rodents Of Unusual Size? I don't think they exist."
—Westley, *The Princess Bride*

In the midst of the darkened jungle, it sniffs the wind and catches the scent of a lone human not more than a mile off. Saliva dripping from its sharp fangs, it eagerly sets off in search of its prey. There is but a sliver of a moon in the sky, but this doesn't matter to the creature's inhuman eyes. The scent grows stronger and the beast slows its movements to a crawl as it silently stalks its prey from the depths of the forest. Then, in a split second, it springs into action. Claws rend flesh in a single swipe. Blood gushes forth. Jaws sink deep into the shoulder, snapping bones as if they were twigs. In an instant, the human is dead.

A lion might not look particularly monstrous while sitting caged in a zoo, but make no mistake, a midnight encounter with one in the wild would change that perception in a hurry. For most people today, there is not much reason to worry about being hunted.* Every now and again the story of a lone backpacker being eaten by a large predator makes its way across the media, but the reality is that predators capable of eating people are mostly endangered and often terrified of even coming close to us.

Yet it was not a long time ago when wild animals were a regular cause of death. In the 1800s, rain forests weren't ecoholiday destinations where tourists could be found snapping photographs of toucans and orangutans. They were jungles where bloodthirsty beasts waited to eat the unwary. Explorers who entered such places often did not come back. And that was in the 1800s. Ancient humans had it much worse.

The Aché people of Paraguay hunt with bows and arrows to this very day and, unlike humans in most other parts of the world who sit alone at the top of the food chain, they are hunted. Jaguars share much territory with the Aché and eat many of the same small mammals that the Aché depend upon to survive. However, jaguars also readily kill the Aché themselves, inflicting an 8 percent mortality rate on males in the population. For comparison, consider the fact that in 2002 the World Health Organization calculated a 2.1 percent mortality rate from road traffic accidents, a 2.2 percent mortality rate from malaria, and a 9.6 percent mortality rate from strokes. It is mind-boggling to think that jaguars could bring a somewhat similar loss of life to males in a population as strokes do in the developed world today, but given that the Aché have the same sorts of rudimentary tools that most of our ancient ancestors had, we have to assume that this was the way life once was.

Yet far beyond the issue of not having advanced equipment for a journey into the wilderness, early humans first exploring the wild had precious little information on what to expect there. One of the

*Until the lunatic owner of an exotic-animal farm in Ohio housing eighteen Bengal tigers, seventeen lions, and numerous other carnivores opens up all of his enclosures to the outside world and then shoots himself in the head.

first encyclopedic works on the natural world was written by the Roman scholar Pliny the Elder, and he wasn't even born until AD 23. Ancient adventurers would have had only tales passed by word of mouth to inform them of what to expect when they stepped beyond the safety of their town or village.

Let's face it, word of mouth distorts, but that was not the only problem. In jungles, dense vegetation blocks most lines of sight and forces visitors to make sense of fleeting glimpses of movement, strange animal calls, and mysterious prints in the mud. There was also the thrill and fear. Adrenaline dramatically alters perceptions. Imagine the first reports: "The beast was as large as a house!" "Its teeth were as long as daggers!" "I once caught a fish *this* big!" You get the idea. It is not hard to see how otherwise ordinary animals transformed into monsters of legend.

With human exploration of the natural world in its infancy, the first environments of mystery encountered were the wild spaces just beyond town. Thus, it is unsurprising that some of the earliest monsters in human history are merely fierce animals with extraordinary characteristics.

Among these was the Nemean lion, a great cat born to Typhon, the godlike creator of monsters, which was later nurtured by the goddess Hera. A description of this beast as a fierce man-eater with skin that could not be harmed by mortal weapons has been attributed to the Greek scholar Apollodorus (180–120 BC). In his account, the monster is hunted by the hero Hercules, who must slay the beast as the first of his great labors. His tale reads: "And having come to Nemea and tracked the lion, he first shot an arrow at him, but when he perceived that the beast was invulnerable, he heaved up his club and made after him. And when the lion took refuge in a cave with two mouths, Hercules built up the one entrance and came in upon the beast through the other, and putting his arm round its neck held it tight till he had choked it."

Exactly how big the lion actually was is not mentioned, but the mythic scene is well depicted on ancient pottery. In these works, the Nemean lion is shown to be as large as Hercules or slightly larger.

Hercules was part god and known for being big and strong. So a lion matching him in size would, presumably, have been large too, but only slightly larger than lions that are alive today. To say that the lion was a giant would be wrong. It is never depicted as dwarfing either Hercules or any humans, so we can assume it was merely meant to be a mean and mostly invulnerable beast.

Herakles and the Nemean Lion, attributed to Kleophrades painter. Greek ceramic stamnos, c. 490 BC. University Museum, University of Pennsylvania, Philadelphia.

But some monsters actually were giants. Homer, who lived around 850 BC, recounted a tale in the *Iliad* of a fierce boar that was unleashed upon humanity. According to the story, there was a king of a Greek city known as Calydon. He was a good king who looked after his people by making frequent sacrifices to the gods, but at one point he failed to properly honor the goddess of the hunt, Artemis. She became angry at this lack of respect and, in a temper tantrum, let her personal boar run wild in the king's lands.* Homer describes this boar: "The Lady of Arrows sent upon them the fierce wild boar with the shining teeth, who after the way of his kind did much evil to the orchards of Oineus. For he ripped up whole tall trees from the ground and scattered them headlong roots and all, even to the very flowers of the orchard."

"Shining teeth" indicates that its tusks were large and sticking well out of its mouth, and to be able to uproot tall trees, it must have been enormous. Artwork supports this last claim with Greek pottery revealing a boar as long as more than two men were tall, suggesting that it would have been about a length of 11 feet (3.4 meters).

The Calydonian Boar Hunt, painted by Kleitias. Greek ceramic krater, c. 570 BC. Archaeological Museum, Florence. Scala/Art Resource, NY.

*When you are the goddess of the hunt, I guess you end up with some pretty exotic pets.

As monsters go, the Calydonian boar and the Nemean lion did not require much creativity. One was just a somewhat large lion with seemingly weapon-deflective skin, the other just a boar that differed from normal animals in its size, strength, and ferocity. All storytellers needed to do was point to the real lions and boars that most Greeks were familiar with and say, "The boar that Artemis released was like that, only bigger and meaner," or "See that lion? The one that haunted Nemea looked the same, but its skin was invulnerable."

And while it might not be easy for modern audiences to appreciate why such creatures would have been frightening, consider this: Male wild boars frequently grow to 5 feet (1.5 meters) in length, weigh over 440 pounds (200 kilograms), and their tusks are more than 4 inches (10 centimeters) long. Today, fatalities from boar attacks in Europe are extremely rare, but traumatic wounds are not. The wounds can be easily treated in hospitals, but a few thousand years ago blood loss and infection would cause many encounters with these animals to be lethal. Boars are exceedingly territorial, and before their habitats were significantly reduced, they were a serious menace. With fear of wild boars already present among ancient populations, it would not have taken much for a mysterious call heard in the woods and unexplained fallen trees to be tethered to the presence of a boar of mythic proportions.

As for lions, they were not just the inhabitants of the African savannas. European lions lived in and around ancient Greece.* To the best of modern scientific knowledge, these were as fierce and as large as the African lions of today. They probably stayed well away from human settlements, but even so, if any lions hunted humans,

*In 2011, a wonderful little article in the academic journal *Biotropica* suggested that a solution to Greece's financial problems might involve reforesting large swaths of land and reintroducing lions to the landscape to encourage ecotourism. While no action on this matter has yet been taken by the Greek government, they might be wise to look beyond ecotourism and consider connecting such conservation actions toward their long and well-known mythical traditions. "Come to Nemea and see the legendary lion that battled Hercules!"

the stories of such events would have spread like wildfire along trade routes and could have quickly led to the imagining of a terrible beast of legendary size. And the tendency for lions to hunt at night would have strengthened the myth. People living near lion dens would vanish in the night, leaving behind just a smearing of bloody remains. Many lion attacks would have had no eyewitnesses to describe what exactly did the killing. And even if there were witnesses, their eyes would have been mostly useless in the dark. They would have seen glimpses of action, heard roars and screams, and been overwhelmed by fear. An accurate report of what sort of predator had attacked would have been impossible. So the stage was set for normal lions to be transformed into supernatural monsters. This is probably where the concept of invulnerability set in. European lions would have been able to survive a number of wounds before being killed. It is not unlikely that watching a lion continue an attack after being stabbed or shot by an arrow led to the rumor that it was impervious to mortal weapons.* Even so, could the ancients have actually been right? Could there have really been a lion or boar of mythic proportions?

Larger than life

Many land animals can get really big. The African elephant can reach over 25 feet (7.5 meters) in length, 11 feet (3.3 meters) in height, and can weigh more than 13,000 pounds (6,000 kilograms). Many dinosaurs were even larger. However, big animals leave behind big bones when they die and tend to be conspicuous in the fossil record. Some of these bones could have been interpreted as the bones of predators, particularly if they were found alongside the skull of an extinct saber-toothed cat or cave bear. It is also reasonable to believe that the fossil tusks of one animal, like those of a mammoth or mast-

*Not to mention unbelievably fierce, since stabbing, but not killing, many large predators can just send them into a furious rage.

odon, were mistaken by the Greeks to be the tusks of a giant boar. Adrienne Mayor's book *The First Fossil Hunters* looks at the interactions between ancient people and fossils, and points out that some of these relatives of the modern elephant were once dwelling on the islands in the Mediterranean and that their remains ended up being labeled as having belonged to heroes, giants, Titans, and even the Cyclops.* There are even some fossil tusks in the Temple of Athena Alea at Tegea that are identified as having belonged to the Calydonian boar. Such discoveries no doubt played a part in the conjuring up of giant animal monsters, but what of living animals? Could a particularly large lion or giant boar species have actually been plaguing ancient humans? In the case of the Nemean lion, the answer is an intriguing yes.

The Eurasian cave lion was similar to the lions of Africa and the recently extinct European lions, but it grew as much as 25 percent larger. And it seems likely that it had interactions with ancient humans since it was alive when people were roaming around much of Europe and Asia. Evidence of the lion does not appear in the fossil record much after 11,000 BC, but numerous cultures that are widely thought to have mastered language and some degree of oral storytelling were already in existence at this time. It is possible that these people captured the essence of this great cat in their tales and passed along stories about it through the generations both to entertain and to warn fellow members of the population about dangerous animals in the surrounding environment.

Moreover, 11,000 BC may not have been the actual date when the cave lion vanished from the face of the earth. Just because nobody has found more recent fossils of the species does not mean it did not linger on. The fossil record is a fickle thing and it often

*The skull of an elephant has a large space in the center where the muscles for the trunk attach. However, trunks are made up of soft tissues that rot away after death. To those without a background in paleontology (a description that pretty much applied to all of the ancient Greeks), such skulls look like they have one giant eye in the center. Moreover, the presence of tusks adds the impression that the Cyclops has fangs.

does not provide a perfect snapshot of ancient history. Fossils form in regions where sediment can quickly fall on top of an animal after it has died. When lions were common in areas near lakes and rivers, their bones would readily get covered up and preserved. Preservation also tends to be good in caves, where lion fossils are often found. However, as humanity spread across the landscape, and freshwater sources along with caves were identified as valuable places to settle, lions would have been driven away from these places and forced into environments where they would be less likely to fossilize after death. They may have died out shortly after being displaced in this way, but it is possible that some small and isolated populations endured for several thousand years longer, particularly in more remote and less hospitable locations at higher altitudes where humans were not commonly spending time. If this was true, then the Mesopotamians of 6000 BC may well have tangled with some very big cats and had more than the stories of their ancestors to give them nightmares.

Moreover, if isolated populations of Eurasian cave lions did hang on in mountainous regions, this may have increased both their size and their resistance to weapons in some rather fascinating ways. To begin with, evolution drives mammal populations dwelling in colder environments to get larger over time. This has to do with the physics of heat. Small bodies have more surface area relative to their overall mass and lose heat from their cores much faster than large bodies. For this reason, big animals are much less resistant to heat loss and tend to better tolerate cold climates. This is why the Kodiak bears living on the frigid Alaskan islands are so much larger than their closely related grizzly bear kin dwelling in milder climates. It is also why the mammals that survived the various ice ages, like the woolly mammoth, giant sloth, and woolly rhinoceros, grew to such enormous sizes. They were extraordinarily resistant to heat loss.

Whether Eurasian cave lions were in fact driven into the mountains and whether natural selection actually did lead to the already large lions getting larger is of course speculation, but the evolutionary mechanisms that connect cold and size are well known, so the possibility is not mere fantasy. The presence of unusually large and powerful lions in remote mountain areas mixed with the adrenaline-influenced perceptions may well have been responsible for people coming to believe in a monstrous lion.

As for the Nemean lion's invulnerability, for weapons to have bounced off of its body, as described in Apollodorus's story, it must have had really tough skin. Is it possible that a lion had such skin?

There is a genetic disease called scleroderma that arises in humans and makes the skin thicken. It is, however, a disease. Those who suffer from it often develop kidney and lung problems that cause a lot of pain. Other animals can develop this condition, but it is hard to imagine scleroderma causing the skin of a lion to thicken to a point where it could deflect an arrow or a sword. Humans who suffer from scleroderma are more fragile than they are invulnerable. Thus, scleroderma as an explanation for the weapon-deflecting properties of the Nemean lion does not make sense. For its skin to have been able to do what Apollodorus said it could, the skin would have had to have been as thick as that of an armadillo.

Could some mutant lion with skin as thick as an armadillo have been around in ancient Greece? Anything is possible when it comes to genetic mutation, where the DNA in an animal spontaneously changes and leads to the expression of physical characteristics that are distinctly different from those of others, but such a lion would hardly have been a threat. A lion with thick armadillolike skin would barely be able to move, let alone hunt and kill humans. So a big, fierce lion that was capable of dodging arrows and shrugging off injuries seems plausible, but a lion with skin that could deflect weapons is hard to believe. However, fur raises some intriguing possibilities.

Just as animals with larger size tend to thrive in cold environments, so too do animals with thicker fur. This is the reason why woolly mammoths dominated much of the landscape during the most recent ice age, and why mammals dwelling in cold environments have much heavier coats than mammals of the same species dwelling in warmer climates. Fur does not fossilize well, so it is hard to know if the Eurasian cave lion had a particularly thick coat to begin with, but if it was driven into cooler regions and forced to endure the cold for a few thousand years, individuals in the population with the thickest and densest fur should have been selected for, leading to the evolution of a great cat with a very thick coat. Of course, a very thick coat of fur can't stop a bullet from a high-powered rifle, but for ancient human hunters wielding bronze daggers and wooden spears, such a coat may have afforded just enough protection to make the lions seem impervious.*

But what of the Calydonian boar? Could a boar of such size have actually existed? Unlike the Nemean lion, which might have been the result of ancient humans tangling with a large, but now extinct, species, no boar species of particularly large size seems to have existed during recent human history. Even so, there is the possibility of a truly giant boar having existed if mutation is considered. In theory, there could have been a small population of mutant boars that happened to live near ancient Greece and grew to twice the size of all other normal boars.†

Regardless of species, all individual animals are limited by the genes that are given to them by their parents. The mixing of genetic material that takes place when sperm and egg unite creates variety, with maternal and paternal characteristics getting shuffled together

*It is also possible that hunters went out to kill a lion, failed, nearly got eaten in the process, and just made up the story of the lion being invulnerable to save face. Adrenaline makes men do and say the stupidest things.

†In 2004, a boar was shot and killed in Alabama that was initially claimed by the hunter to be 12 feet (3.6 meters) long and to weigh more than 1,000 pounds (450 kilograms). This was more than twice the length and weight of the wild boars known to be roaming the U.S. wilderness, so a team of researchers funded by the National Geographic Society went to check out the grave of this creature that locals were calling "Hogzilla." As mentioned earlier, the adrenaline of the hunt leads to mistaken perceptions, and the hunter was—no surprise—proved to have exaggerated the animal's size. When the researchers dug up the animal, they found the boar to be only 7 feet (2 meters) long and weighing in at 800 pounds (360 kilograms). That's a big boar, but still very close to the upper range of natural wild boars. Hogzilla was thus something of a hoax.

into a novel mix, but the mixing does not lead to the birth of offspring that have traits totally unlike any seen in the parents or extended family. Take human families as an example: Two tall parents are likely to have tall kids. Some of these kids might even ultimately grow taller than their parents, but only by inches. It would be preposterous for a child to grow to twice the size of his parents, unless he was a mutant.

This discussion may sound like it belongs in a comic book with optic blasts, adamantium, and superhuman healing, but mutants are very real and they are among us.

The pituitary gland, located at the base of the brain, is responsible for releasing hormones that control various functions of the body, including growth. Growth hormones are critical for human development, but too much of them circulating in the blood can lead to abnormal growth. Normally, the pituitary does exactly what it is supposed to and releases growth hormones in strictly controlled levels. However, sometimes benign tumors develop in the pituitary gland and cause it to secrete higher growth hormone levels than it is supposed to. In most cases, these tumors develop during adulthood after bones in the body have stopped growing in length. In these situations, the released hormones cause the forehead to become more prominent, the jaw to thicken, and hands and feet to become steadily larger, leading to the disease known as acromegaly. The body becomes deformed and more robust, but it does not become significantly longer or taller. However, in some uncommon cases, growth-hormone-secreting tumors in the pituitary can develop during childhood before the body has stopped growing. This early tumor growth leads to increased growth hormone levels, which drive the bones to grow much more than they normally would. In most of these cases, this situation can be made even worse if puberty is delayed or disrupted due to the effect of the growing tumor on the other cells of the pituitary gland regulating puberty. Under these circumstances, the bones just keep on growing and people can end up at the extreme height of up to 9 feet (2.8 meters). Most intriguingly, it is possible for these tumors to be associated with a mutant gene that can be passed along from parent to child, meaning that populations of very large people can emerge in locations where these genes are common.

Such musings are not mere speculation. Charles Byrne, the Irish Giant, who was 7 feet 7 inches (2.3 meters) tall, left Ireland for England in the mid-1700s to make a career on the carnival circuit. He was initially thought to be an isolated individual. When he died in London in 1783, researchers kept his skeleton in a museum and speculated about the cause of his disease. In 1909, with the discovery of the pituitary gland tumors that can cause gigantism, Byrne's skull was cut open, and it was found that he had suffered from exactly such a tumor. Another hundred years later, after seeing a historic picture of Byrne standing next to two twin giants from a nearby village in Ireland, a team led by Márta Korbonits at the London School of Medicine started speculating that perhaps Byrne was not an isolated rarity after all. "There were a lot of stories, folktales, and names of places, hills related to giants exactly in the region where Byrne was coming from," explains Korbonits.

Keen to look at this more closely, Korbonits and her team collected DNA from one of Byrne's molars, which had been preserved for more than two hundred years in a museum. They analyzed the DNA for the possible presence of a mutant gene. Remarkably, they found something.

In 2011, they explained in the *New England Journal of Medicine* that Byrne did indeed have a genetic mutation predisposing him to develop a pituitary tumor leading to his extreme height. Moreover, when they analyzed extremely tall, living patients from the region where Byrne had been born 250 years earlier, they found they also carried the mutant gene. Interestingly, some people carry the abnormal gene but never develop a pituitary tumor. Why the gene causes only some people to become giants is still unknown, but the results make it clear that giant humans have existed throughout history, and while giants themselves are often sterile and cannot give birth to more giants, their siblings who carry the gene, but who never develop the tumor, have the genetic potential to create concentrated clusters of giants by having many children.

Yet it is crucial to recognize that real giants are sufferers of a disease. "Patients with adult or childhood-onset pituitary tumors suffer from many problems and die young if not treated. They frequently have excruciating headaches, go blind, have joint pains, develop high blood pressure, diabetes, heart problems, and lung disease and almost always suffer severely from the change in their appearance and the physiological burden of this," explains Korbonits. Fortunately, today these patients can be recognized early, partially because the genetics of their families are becoming well identified, and the tumors are destroyed before they are able to transform people. "My motto is, 'No more giants,'" says Korbonits.

So if this is possible in humans, is it possible in other animals? Nobody is really sure. The evidence so far seems to suggest not. Although giant humans are rare, they are common enough in the population for us to become aware of them, as we did with Byrne and his Irish kin. If the same condition occurred in other animals, people would notice. Veterinary scientists have documented a fair number of dogs and cats that have pituitary tumors, but these tumors all lead to the sorts of characteristics that are similar to those seen in humans who develop the tumors after puberty. Thick, folded skin and distorted skulls develop, but gigantic size does not. For a giant dog or cat to exist in this way, there would need to be puppy- or kitten-onset of the pituitary tumor and this does not seem to exist in the veterinary literature.

Even if a giant boar could result from piglet-onset acromegaly, experts doubt that it could behave in any sort of threatening way because physiology and body mechanics change with size. Zoologist Steven Thompson at the Lincoln Park Zoo in Chicago argues that it would be unlikely for an abnormally giant boar to be very ferocious or fast because of scaling. "The structure of joints, limbs, leverage, and tensile strength would be entirely out of whack. While an acromegaliac animal may, in principle, seem fearsome, they would be more likely to be slow and awkward," he says.

All science aside, there is also a more practical problem with the concept of a giant boar having ever existed. Let's face it, if a hunter at any point in the past three thousand years picked off a boar twice the size of any other boar ever seen, the bones of that giant animal would have been preserved, put on display, and written about.* An argument could be made that boars are too rare today and that their populations are not big enough to have the genetic diversity necessary for the appearance of a giant member of their species to be statistically likely to arise from mutation. This is a fair point, but what about other animals, like foxes, goats, and coyotes? These animals all have huge populations, yet try finding any giant versions of them in museum exhibits. There aren't any. With such huge populations, it would be expected that at least a few giants would turn up every now and again and be displayed or sold as oddities, but they do not. Either way, the possibility of a giant mutant boar or a small population of giant mutant boars having once existed looks doubtful. Instead, Thompson suggests that if the Calydonian boar is connected to the sightings of a real animal, it would be just a very large specimen of a normal animal. "I'm thinking we could be talking about an NFL defensive lineman of the boar population, an animal that naturally reaches the upper spectrum of large size and ferocity by eating well and learning, through various encounters, to behave aggressively," he explains.

Regardless of how they came to exist as monsters, both the Calydonian boar and Nemean lion lost their monster status over time. The lion appears almost exclusively as a skin worn by Hercules, with the mouth forming a sort of hood over his head in later art, while the boar is always vastly outnumbered by hunters with spears. This is a pretty substantial departure from early artwork. Moreover, Renaissance portrayals of these monsters are not particularly scary, which is surprising because they easily could have been.

During the early and mid-1500s in Europe, artists like Leonardo da Vinci, Giuseppe Arcimboldo, and Albrecht Dürer were demonstrating a tremendous understanding of how to control light and shadow in their paintings. By the time Peter Paul Rubens painted his version of the Calydonian boar hunt in 1611, he certainly knew

*Tabloids . . . not a modern invention.

how to create a creepy or frightening scene if he wanted to. He had the artistic ability to put the boar in a dark forest full of shadows or have the monstrous animal charging out of a thicket hell-bent on destruction. Instead, his rendition of the hunt is brightly lit, colorful, and the boar is placed standing still in the foreground, as if waiting to be stabbed. If it were not for all the spears in everyone's hands and the well-known mythology, viewers could have missed the fact that the painting was about a monster at all.* Yet it is hardly surprising that such a painting was made at this time.

The Calydonian Boar Hunt, by Peter Paul Rubens. Oil on panel, 59.2 x 89.7 cm., c. 1611–1612. J. Paul Getty Museum.

During the centuries after the days of the ancient Greeks, as the human population grew, forests became better explored and animals started being properly documented. Communities sprouted up in previously unpopulated areas, and towns grew larger. The natural world became less scary, and many predators quickly became the hunted as weapons improved. If a giant boar or lion species had ever actu-

*And named it something ridiculous like *Picnic with Boar.*

ally existed, it had to have been driven to extinction. Certainly the normal-sized European lion, which may have inspired much of the fear that fueled the Nemean lion legend by simply living near Greek communities, went extinct around 100 AD. Fewer and fewer lions and boars of any size would have been around, and as they faded from the landscape, so too did monstrous stories associated with them. But even with the passing of the Calydonian boar and Nemean lion, monstrous animals did not cease to exist. A new threat emerged in the stories of Persia and Arabia around 1300 AD: the Rukh.

Feathery death

Most famously described in the popular Persian folktales of Sinbad the sailor, as translated by Sir Richard Burton, the presence of the Rukh is first revealed by its egg. Sinbad, stranded on a newly discovered and seemingly uninhabited island, cannot work out what the giant white dome is when he first spots it. As he walks closer for a look, the summer day suddenly goes cool and the sky goes dark. Sinbad figures it has to be a cloud. "Methought a cloud had come over the sun, but it was the season of summer, so I marveled at this and, lifting my head, looked steadfastly at the sky, when I saw that the cloud was none other than an enormous bird, of gigantic girth and inordinately wide of wing, which as it flew through the air veiled the sun and hid it from the island." The giant bird then comes in for a landing, settles on the dome, and begins brooding the egg.

Sinbad, desperate to get off the island, unties his turban and uses it as a rope to hitch a ride on the Rukh when it flies off the next morning. It takes him to another island, where it first attacks and catches a large snake and then a rhinoceros. Later in his voyages, the Rukh attacks Sinbad and his fellow travelers after they have broken open its precious egg.

Sightings of the Rukh are not limited to the tales of Sinbad. Marco Polo supposedly saw a bird so large that it could carry off an elephant in its talons and then drop it to its death from high above.

There is no possibility of a bird having ever existed that could fly off with an elephant in its talons. This is not a mere matter of paleontology having failed to turn up the bones of such a beast. There is no need even to go searching for potential Rukh fossils, because the laws of physics get in the way.* For the Rukh to have been able to have carried what people say it carried, it would have needed to have had a wingspan greater than 50 meters (264 feet). To put that in perspective, such a wingspan is as long as the largest dinosaurs (which were themselves larger than many office buildings). It is five times larger than the wingspan of the largest known flying creature, a pterosaur from the age of the dinosaurs known as *Quetzalcoatlus northropi,* which seems to have already been pushing the boundaries of flight physiology and is widely thought to have behaved like a vulture that ate dead dinosaurs or hunted for prey while on the ground

---

*The physics supporting all of this is a chore, but if we work with the assumption that, like eagles, the Rukh could lift half its own body weight into the air to fly off with a 4,500-kilogram elephant, the bird would be around 9,000 kilograms, and the combined bird plus elephant payload would be 13,500 kilograms. Classic lift theory states that: Lift = $\frac{1}{2}(p \bullet v^2) \bullet S \bullet Cl$, where p is the air density (1.18 kg/m³), v is the velocity (say, 9 m/s), S is the projected area (m²), and Cl is the dimensionless lift coefficient (say, 1.15, based on measurements for a vulture). This means the bird would need a wing area of 245 m² to stay aloft carrying an elephant. In aerodynamics, the aspect ratio of a wing is essentially the ratio of its length to its breadth, a measurement called the chord. A high aspect ratio indicates long, narrow wings, whereas a low aspect ratio indicates short, stubby wings. For most wings, the length of the chord is not a constant but varies along the wing, so the aspect ratio AR is defined as the square of the wingspan b divided by the area S of the wing platform. This is equal to the length-to-breadth ratio for constant breadth:

$$AR = \frac{b^2}{S}$$

Bird aspect ratios can vary from 1.5 to around 18, so let us assume something around 10. Therefore, a 245 m² wing area would mean the Rukh would need a wingspan of around 50 meters. Of course, there is a problem here, because this applies only to a Rukh that is already in flight. To behave as legends say it did, with such a "small" wingspan it would need to swoop down at high speed, grab an elephant, and fly off without losing momentum. If the bird were to stop, kill the elephant, and then try to get lift again with the elephant in tow, such wings would not have worked at all. If you want to understand the physics of that sort of behavior, get a degree in aeronautical engineering.

like marabou storks rather than carry prey anywhere. The concept of such a large bird presents numerous physiological problems, like how it could have had a heart large enough to pump blood out to its wings and how its bones would not have broken under its own weight.

So if such a monster could not have actually existed, where did the idea of a giant bird of prey come from? The largest birds alive today are the Andean condors. They are huge, with wingspans that sometimes extend as far as 10 feet (3 meters) in length. But they are docile scavengers that simply soar along the edges of canyons and over plains in search of carrion; picking up and dropping prey is not part of their repertoire. Moreover, they are found only in South America and cannot be associated with the Rukh legends, who created these tales, since the Persians and Arabians had not made it to the New World yet. There are some birds of prey that do feed on reasonably large mammals. The harpy eagle plucks monkeys out of trees, but it too lives only in South America. However, the crowned eagle in sub-Saharan Africa is well documented for behaving in a similar fashion. It feeds primarily on medium-sized mammals. And if monkeys and small antelope are on the menu, one has to ask, why not small humans?

Modern predatory birds are perfectly capable of causing people a lot of harm. There is one particularly frightening description in the *New York Times* in 1895 of two boys in California being fiercely attacked by an eagle as they approached its nest. Neither died, but one was disfigured and blinded. In a similar vein, in 2008, a boy in Michigan was hospitalized after an eagle attacked when he approached its nest. However, as much as eagles can hurt people foolish enough to get too close, they are not known to actively seek out humans for food. But this might not have always been the case.

On the South Island of New Zealand, there once was a bird of prey larger than a harpy eagle. Known as Haast's eagle, this bird probably weighed up to 33 pounds (15 kilograms) and had a wingspan of around 10 feet (3 meters). While hardly a giant, its diet is widely thought to have consisted of the flightless birds known as moas, which were often the size of adult humans. Haast's eagle lived on the island undisturbed until people arrived and started eating all the moas they could find. With no food to eat, the eagle went extinct. Exactly when this occurred is a mystery. Many records put its extinction date during the 1400s; others propose that it survived into the 1800s. Regardless, this was a bird capable of attacking and eating animals that were sometimes larger than humans, and it does not take much imagination to envision early islanders being killed by hungry Haast's eagles.*

Another recently extinct bird that may have fueled the Rukh legend is the elephant bird of Madagascar, *Aepyornis,* which survived until 1030. It was huge, 10 feet (3 meters) tall, but hardly threatening. It was flightless, herbivorous, defenseless, and did not survive for long once humans started inhabiting the island shortly after 500. While they probably tasted like chicken and scared nobody at all, the discovery of such large birds may have led to stories that these were the not-yet-fledged juveniles of a much larger predatory bird.

However, when it comes to the Rukh legend, *Aepyornis* and Haast's eagle present problems of timing. Madagascar was not discovered by the Europeans and Arabs until the 1500s, and New Zealand was discovered by these groups only in the 1700s, but the Rukh starts being mentioned in stories hundreds of years earlier. It seems reasonable enough that the discovery of these birds may have increased belief in a monster that was already alive in the minds of

*For years, archaeologists assumed that punctures found in the skulls of early human ancestors, like some of the australopithecines living 3.5 million years ago, were made by the fangs of great cats, but in 2006 a team of a researchers published evidence in the *Physical Journal of Anthropology* revealing that these punctures were nearly identical to punctures that large eagles make with their beaks and talons when they kill monkeys today. This led the team to argue that birds of prey were playing a big part in hunting our forebears. Such finds also raise the question of whether we might have some instinctive fear of large raptors buried deep within our genes.

sailors, but it begs the question of where the idea for a giant bird that was flying off with elephants initially came from.

Part of the legend of this monster undoubtedly came from people encountering fossilized dinosaur footprints. While dinosaurs like *Triceratops, Stegosaurus,* and *Diplodocus* left footprints that could never be misinterpreted as having belonged to birds, the dinosaur group *Theropoda* that *Velociraptor, Tyrannosaurus,* and many other carnivores belonged to had members with feet that were distinctly birdlike. Did someone find these sorts of tracks and conclude that a giant bird once walked by? It seems plausible, but what about the idea of elephant-dropping behavior?

In *The Travels of Marco Polo,* Marco Polo wrote: "It is so strong that it will seize an elephant in its talons and carry him high into the air, and drop him so that he is smashed to pieces; having so killed him, the bird gryphon swoops down on him and eats him at leisure. The people of the isle call the bird Ruc."

What is astonishing is that this description precisely mirrors the behaviors of the bearded vulture, a bird dwelling in Africa and Asia today. It uses a tactic of grabbing the large bones of recently dead animals, flying high, and then dropping the bones onto rocky terrain below. The impact breaks the bones into pieces and allows the vultures to feed upon the juicy marrow inside.

That the Rukh seems to have done exactly what bearded vultures were doing, but on a grander scale, hints that the formation of this monster actually *required* the careful observations of the natural world that ultimately played a part in driving the Calydonian boar and Nemean lion to their end.

But why invent a Rukh at all? To answer this, it is helpful to look at what the Rukh offered that both the Calydonian boar and the Nemean lion did not. At its core, the Rukh differs from the boar and lion only in its ability to fly. Neither the boar nor the lion could contribute to the fears of ancient sailors heading off into the unexplored seas. The Rukh, on the other hand, because it had wings and could soar over oceans, remained a threat.

In Hindu mythology, the Rukh had something of a kindhearted

alter ego in the form of Garuda, a giant bird of prey that the god Vishnu rode as an aerial mount. Garuda was a champion of good and frequently depicted in the *Mahabharata* as hunting venomous snakes and snakelike creatures. And in modern mythology, J. R. R. Tolkien presents the giant eagles in *The Lord of the Rings* as allies who save the wizard Gandalf when he is imprisoned by the villainous wizard Saruman. Why such variation?

It is impossible to know what the inventors of these benevolent beasts were thinking when they created them, but one possibility might be that regional understandings of birds and snakes differed between those dwelling in Middle Eastern settlements where the Rukh legend formed and Southeast Asian settlements where Garuda did.

Certainly, from the perspective of snakebites, these regions show very different patterns. In 1998, a review of the morbidity and mortality of snakebites in locations around the world was published in the *Bulletin of the World Health Organization*. Aside from pointing out that deaths from snakebites are a major health problem in many places, the report presents some of the best available data on where humans are most likely to get bitten by a snake, fall ill from the bite, and die.

In India, the situation is relatively bad. The number of bites per year that actually get reported averages around 114.5 per 100,000 people, and roughly half require medical intervention to treat the spread of the venom through the body. Fortunately, due to the availability of antivenins, only 3 people per 100,000 typically die annually. In the Middle East, the situation is staggeringly different. An average of 12.5 bites are reported per 100,000 people and only 0.062 people per 100,000 die.

In fairness, perhaps it is not appropriate to look at current deaths from snakebites, since this could represent more of a commentary on the state of medical facilities in these two regions than on the threats actually presented by snakes, but the sheer number of snakebites being reported per year is probably a somewhat fair representation of the overall snake threat to the local population. With this in mind, it is pretty obvious that the snakes of India are a much more serious hazard than the snakes of the Middle East. And a 1954 report in the *Bulletin of the World Health Organization* confirms this by concluding that Southeast Asia was the fatal snakebite capital of the world at the time.

In both regions, there are birds of prey that hunt snakes, but among locals in India, where snakes currently inflict nine times more bites per 100,000 people than they do in the Middle East, one has to wonder if seeing birds of prey killing these dangerous animals might have been viewed as something positive. Did the goodness of Garuda have to do with the serious dangers presented by snakes? Was it the snake-hunting services provided by birds in the region that led to the invention of the noble Garuda in the first place? This all seems likely.

As for Tolkien, he was writing at a time (the 1950s) when it was well known that birds of prey attack snakes far more often than they attack people. Whether this knowledge was what guided his decision to present giant eagles as allies rather than enemies is unknown, though. What we do know is that giant animals have continued to exist as monsters all the way up to the present day.

An end to mythic proportions?

Many modern giant animals are not taken seriously. Indeed, in Rob Reiner's 1987 film *The Princess Bride,* as the protagonists wander their way through the perilous fire swamp, the princess turns to her protector and asks, "Westley, what of the R.O.U.S.'s?" He responds, "Rodents Of Unusual Size? I don't think they exist." Westley is then, of course, promptly attacked by a giant rodent.

This scene is not intended to frighten viewers. The giant rodent isn't realistic, believable, or scary. When it yelps in pain, it sounds more like a Muppet than a monster. The humor in the scene stems from mockery because, to modern audiences, animals of mythic proportion seem totally absurd.

Yet these monsters are not as extinct as they might at first seem.

Giant boars and huge birds of prey may no longer feature as monsters in modern culture, but many other animals do. Consider the birds that feature in Alfred Hitchcock's *The Birds* or the spiders that are the stars in Frank Marshall's *Arachnophobia.*

Here are animals that, like the Nemean lion, are tweaked in some very basic way. Hitchcock's birds look like normal birds but behave like crazed predators, hungering for human flesh. Marshall's spiders are different from other spiders only in that they are highly resistant to insecticides. Indeed, the really scary aspects of the monsters in these movies—the fact that the birds swarm people and that the bite of the spiders is lethal—are very much real. Birds can and do swarm people and there are spiders that have bites so lethal that they can kill people quickly.*

Once again, study of the natural world is responsible for the birth of these monsters. The idea of birds killing by swarming or spiders wiping out an entire village with their venomous bites could have come about only as a result of people seeing such things happen in the real world. And it is precisely because these monsters are so easily accepted by educated modern audiences that they have succeeded in striking so much terror into our hearts.

So animals still exist as monsters, but instead of scaring people with their size and strength, they now do so with natural abilities that are subtly altered by creative Hollywood minds to be more malevolent and threatening than they actually are. Yet there is, of course, one monstrous animal that is an exception. Hardly small and venomous, or even remotely in the same category as any other modern animal monsters, is the colossal ape, Kong.

---

*Seagulls can be downright vicious. Accidentally wandering into their nesting sites will almost always lead to a flurry of feathers, shrieks, mobbing, and pecking. Moreover, a study published in *Nature Geoscience* in 2011 found a period of bizarre seabird behavior in 1961 when seagulls frequently slammed themselves into beach homes and cars (widely thought to have inspired Hitchcock's film), which came about as a result of the birds suffering nerve damage after being exposed to neurotoxins released by a toxic algae bloom off the coast of California.

A king among the giants

Created for the cinema in 1933 by Merian Cooper and Ernest Schoedsack, Kong has since been reincarnated for *King Kong* movie viewers nine times. In every single film, Kong is found in a remote part of the world that has remained unexplored for one reason or another. These remote places are always home to other huge animals; in some cases these are dinosaurs, in others, they are just very large versions of common animals like snakes and lizards. The story line for each film brings humans to Kong for different reasons, but the ape always develops an emotional bond with the leading lady, nurtures and cares for her in the wilderness, and ends up being shot by aircraft while trying to protect her from harm.

Yes, Kong is big, frightening, and lethal. In every film dozens of humans are slain by Kong's actions, but he is not specifically revealed as malevolent as the man-eating Nemean lion or the orchard-ravaging Calydonian boar. Moreover, all of the films end with a sense of sadness over the death of the giant ape. There is no feeling of terror or fear of Kong by the end of the films, only sympathy.

At first glance, Kong's story seems the same as that of other monstrous animals. Like the Nemean lion, he is a monstrous animal from unexplored wilderness wreaking havoc on civilization. But Kong is different from these monsters of Greece. Ancient monstrous animals are not natural. They are the products of the gods and are sent to Earth to create pain and suffering. The Rukh, on the other hand, is not brought to the world by gods but discovered by humans; it just attacks those who tamper with its eggs. The spiders in *Arachnophobia* are dangerous animals from the deep jungle that are accidentally brought by humans back to civilization. Kong also is a dangerous natural animal brought from the wilderness to civilization but he is intentionally brought back for profit. Really, Kong's story is a morality tale about the consequences of exploiting the natural world, raising the question of whether it would have been better to have just left Kong alone.

Most intriguingly, Kong is distinctly a "he," and that cannot just be dismissed. It indicates a key difference in the way audiences identify with the monster. None of the other giant animal monsters are ever described by gender. By being given gender, Kong is, in a way, brought one step closer to humanity and made less of a monster. In our quest to understand what monsters say about human fears through the ages, such a dramatic difference between the most famous modern giant animal monster and ancient monsters of the same type is intriguing. Why the decision to make the men in the story so insensitive and cruel? Is there something in society that we now fear more than the jungle?

2

Beastly Blends—Chimera, Griffon, Cockatrice, Sphinx

"These creatures you have seen are animals carven and wrought into new shapes."

—Dr. Moreau, *The Island of Dr. Moreau*

Not far removed from the realm of the Nemean lion, Calydonian boar, and Rukh are monstrous animals not of unusual size but of unusual form. Seemingly inspired by ancient people dwelling in a drug-induced haze, many of these monsters are bizarre anatomical blends that have been unbelievably stitched together. Multiple heads, duplicated limbs, bodies combining wings and claws, and tails with teeth, these monsters proved common throughout most of ancient and medieval history.

Among the most notable are the Manticore of Persian myth, a beast that had the head of a human, the body of a lion, and a tail covered in venomous spikes that it could launch at its foes like arrows. There were also the many-headed monsters, like the reptilian Hydra that battled Hercules, and the three-headed hound

Cerberus who guarded the gates of Hades. And some blends just mixed and matched animal traits without any sort of logic, like the Griffin, which had the body of a lion and the head and wings of an eagle, and the Cockatrice, which was part rooster and part dragon. Yet few beastly blends are as well known today as Chimera of Lycia.

First mentioned in Homer's *Iliad,* Chimera is described as "a thing of immortal make, not human, lion-fronted and snake behind, a goat in the middle." As if being part lion and part goat with a snake for a tail were not disturbing enough, Homer added that it had the ability to breathe a "terrible flame of bright fire" as well.

Some descriptions were far bolder. Hesiod suggested in his *Theogony* that Chimera had multiple heads, one of a "grim-eyed lion," another of a goat, and a third of a serpent, which some translations describe as a dragon. As for its body, its front portion was leonine, the hind portion reptilian, and the torso goatlike. Hesiod, like Homer, suggests that the monster was capable of breathing fire, but contrary to what modern audiences might expect, it is the goat rather than the dragon head that seems to have done the fire breathing.

Chimera of Arezzo. Bronze, Etruscan, c. 400 BC. Museo Archeologico Nazionale, Florence. Art Resource, NY.

Ancient artists went wild with Chimera. Some portrayed it as Homer described, with a snake for a tail, a goat's torso, and a lion head with lion forelimbs, while others followed Hesiod's description, placing three heads on the front of the beast, with lion forelimbs, reptilian hind limbs, and, occasionally, dragonlike wings sticking out from its back. Some artistic representations of the monster fall in between the two, portraying a beast with a goat head awkwardly stuck out of the creature's spine with reptilian and leonine traits mingled throughout the rest of the body.

Yet as often as Chimera appears in Greek art and literature, the fears that it represented are not as straightforward as those presented by the merely monstrous animals discussed earlier. It seems logical that Chimera partially stood for the unknown dangers beyond civilized lands since it was said to dwell in the land of Lycia, around 500 miles east of Athens, in what today is southeastern Turkey. Even so, the dangers posed by wild beasts were well presented in more mundane monsters like the Nemean lion and Calydonian boar. Chimera had much more to it, but why? What was it that drove people to fear a fire-breathing creature with such a blend of unusual animal traits?

Morphological mess

It is a challenge to conceive of the possibility that Chimera's creation was the result of a misinterpreted animal encounter. It just does not make sense. Female lions work in groups to attack prey at night, so it would not be unlikely for a survivor to come away claiming to have been attacked by a lion with multiple heads. But with a snake for a tail?

Snakes are not known for teamwork, even within their own species, and the idea of a snake being associated with a lion attack is hard to believe. It is conceivable that a poor soul might have been ambushed by both a viper and a lion at the same time during some unlucky evening. But it is a stretch to conceive that this was common enough to create a myth.

And then there is the goat. Who came up with this? Goats are eaten by lions.* There is no getting around it: The concept of an ani-

*And in some regions by snakes too!

mal encounter creating Chimera just doesn't make sense. So if the idea of ecology inspiring Chimera is illogical, could there have been something biological behind it? Like other animal-inspired monsters, it is worth considering whether a mutant of some sort was born that the ancients viewed as a monster.

Animals can sometimes be born with the traits of other animals on their bodies. Horses, which normally have just one toe, can be born with multiple toes, dolphins can be born with legs, and humans can be born with tails. These mutations happen when very old genes, which have been inactive in an animal for years, for reasons not entirely understood suddenly become expressed instead of the genes that normally should. Called atavisms, these genetic mutations can create chimeric-looking organisms.

But atavisms are not totally random in what they do. They do not simply grant a random trait. Humans are not born with wings and snakes are not born with fur. Instead, they produce traits that existed in the evolutionary past of that animal. Think about it: Human ancestors had tails, horse ancestors had multiple toes, dolphin ancestors had limbs. Odd as it is, there is a logic to the mutation.

Yet Chimera does not show characteristics of lion ancestors mixed with those of a lion; it has characteristics of a goat and snake. This is not helpful. If the lion were the base animal that an atavism were taking place in, the lion could be born with a weasel-like face and shorter limbs because the evolutionary lineage that eventually led to lions contained a lot of weasel-like predators. But it would not be possible for a lion to be born with any snakelike traits because lions did not evolve from snakes or their relatives. The same is true of goats. Lions and goats are actually relatively closely related, but traits that are identifiably goatlike, like a goat head with horns, appear in the goat evolutionary lineage long after it breaks away from the evolutionary lineage leading to lions.

Even if the evolutionary pathways were different and the lion's ancestry did stem from a lineage that had given rise to goats and snakes, it is hard to imagine a mutant carrying all of these traits surviving birth, let alone its first few months of life, with the sorts of deformities that would have been required for Chimera to have existed. For these reasons it is impossible for an animal looking like Chimera to have ever been born.

Yet the basic ideas upon which Chimera is founded are not, on their own, totally ridiculous. Take, for example, Hesiod's description of the beast having multiple heads. Animals can have two or more heads when twins or triplets develop abnormally in the womb.

Under normal circumstances, an embryo forms from a sperm and an egg. But sometimes, shortly after fertilization, the cluster of cells destined to become the embryo spontaneously splits into two fully functional clusters of cells that go on to become two embryos. This results in identical twins. However, it is speculated that sometimes the two clusters of cells do not entirely break away from one another or that after breaking away they partially reconnect. The result, if the twins survive to birth, is conjoined twins.

Among humans, such individuals' lives can be medically challenging, as two brains are often forced to share vital organs like stomachs and hearts. For animals, survival is unlikely, since obtaining food requires finely tuned senses and considerable coordination. Two brains trying to control a single body makes coordination difficult, and animals with two heads are rarely seen surviving to adulthood in the wild. Even so, two-headed and even three-headed animals* can exist and might have inspired the creation of Chimera and other beastly blends like Hydra and Cerberus.

The idea of having limbs that differ from one another in size and shape is not preposterous either. It is perfectly reasonable for an animal to be born, as the result of mutation, with hind limbs that are shorter than the forelimbs or are deformed and thus different-looking from the rest of the body. Again, these sorts of mutations would make life very hard, and survival to adulthood would be extremely unlikely. Although it is (just barely) feasible that a multiheaded lion

*The result of triplets that end up connected to one another in the womb.

with deformed hind limbs that were mistaken for goat legs could have existed, the poor beast would not have been able to hunt.*

No, mutation seems as unlikely an explanation for the origin of Chimera as a multiple-animal ambush in the woods. But there is another possibility: fossils.

Sticky situation

In *The First Fossil Hunters,* Adrienne Mayor makes a persuasive argument that the half-eagle, half-lion monster known as the Griffin came about when ancient people discovered the bones of the dinosaur *Protoceratops* and tried to imagine what the animal would have looked like when alive. The idea is entirely logical—*Protoceratops* had a beak but was clearly something quite different from a bird. Moreover, *Protoceratops,* like many of the other members of the dinosaur group to which it belonged, like the iconic *Triceratops,* had a bony frill rising up from the back of its neck. This frill is thought by some paleontologists to have played a role in protecting the cervical vertebrae just behind the dinosaur's skull. It is not unlikely that the first people who looked upon this fossil identified the bones of the frill as having been some sort of wing, particularly in light of the fact that there was a beak on the business end of the beast.

The Griffin is not alone in potentially being explained by observations of early fossils. The Cockatrice, which appears in much medieval literature and art, is presented as something of a malevolent chicken blended with reptilian or draconic features that could kill with its gaze or petrify with a touch of its toothy beak. Although a deadly petrifying gaze is one of the characteristics associated with Medusa, the morphology of the monster is strikingly similar to

---

*Honestly, instead of featuring as a monster in Greek mythology, it would have made more sense for such a creature to have played the lead role in one of Sophocles' or Euripides' tragedies.

many fossils dug up in China during the past decade that walk the fine line between being somewhat like modern birds and somewhat like the aggressive predatory *Velociraptor* featured in *Jurassic Park*. The first creature of this sort uncovered by modern paleontologists was an *Archaeopteryx* found in Germany a little over 150 years ago. The existence of the Cockatrice in historic literature and art raises the question of whether somebody somewhere stumbled upon a similar fossil a long time earlier. With such remarkable similarities between the fossil and the Cockatrice, the possibility seems likely. But even so, individual fossils encased in rocks go only so far in explaining monsters that mix various animal traits. It is hard to think of a single extinct species that could explain Chimera, but there are still some possibilities if the fossilization process itself is taken into consideration.

One type of fossilization process that can occur is a catastrophic flood. In desert regions, when parched land is suddenly exposed to intense thunderstorms, a flash flood can result. The rushing water often sweeps away animals in its path. Floodwaters frequently are so violent that the creatures they carry away die, and then, as the flow slows or as the rushing water turns a bend, the corpses of the animals are deposited. It is conceivable that a goat, a lion, and a snake were all killed, dumped in a mangled pile, and then covered up by sediment carried by the remaining waters from the storm. These jumbled bones would be difficult to make sense of and could have been interpreted as having been the bones of a single fantastic monster. Similarly, several wolves killed and dumped by raging waters in a single location could have led to ideas surrounding the multiheaded Cerberus, and several large reptiles captured in this way could have inspired the myth of the Hydra. Yet there is another option.

Most corpses get buried and transformed into fossils in places like lake beds and stream banks where sediment can quickly pile up on top of the corpse shortly after its death. The speed with which this happens is important, because the longer it takes to cover up a corpse with sediment, the greater the chances that a scavenger will come along, tear up the body, and scatter the bones so they are not buried. In general, animal interactions with corpses are bad for potential fossil formation, but there is one very significant exception to this.*

In places where petroleum naturally bubbles up to Earth's surface, pools of tar can form. No sensible animal will go near these pools, since tar is both sticky and smelly, but the thing about tar pits is that they can sometimes get covered up and appear benign. Shortly after a rainstorm, water can flow into tar pits and cover the tar, making it look like a lake. "Alternatively, in some areas dust and dirt can blow over tar pools, plants can begin to grow over the gloopy tar, and this can entice animals near," explains paleontologist Michael Benton at the University of Bristol. In the days or weeks after water has accumulated or plants have grown, a thirsty or hungry animal might wander down and step into the tar. While trying to free itself, the animal only becomes more hopelessly stuck and sinks deeper into the quagmire. It shrieks in panic as it slides inevitably toward its death, but its cries bring the attention only of predators keen to take advantage of what they think is a wounded animal.

With the first animal trapped in the tar just a short jump from the shore, a predator leaps in to take advantage of what looks like easy prey. The predator might get one or two good bites before also becoming stuck. The predator's attacks turn to wails of frustration and terror as it too slides toward its demise. These cries attract more predators, and the scenario continues. Ultimately, a pile of animals forms, some dead, some still alive, and the chaos, along with the scent of dead meat, attracts vultures that come in for a nibble. Some

*Predator/prey interaction fossils are very occasionally discovered, and they are fascinating. A fossil of a *Protoceratops* struggling for its life against an attacking *Velociraptor* was found in Mongolia. The predator has its claws wrapped around the prey and the prey was clearly doing everything it could to throw off its assailant, but the battle ended with both dinosaurs dead, as either a sudden sandstorm quickly buried them or a sand dune collapsed on them as they struggled. It is one of the most impressive fossils ever discovered, mostly because such interactions are exceedingly uncommon in the fossil record. (FYI—You can buy a re-creation of this for $9,500 (plus shipping!). www.bhigr.com/store/product.php?productid=464.)

land, feed, and get away, while others land, get stuck, and die. Eventually, the process ends when the entire mangled and smelly mess sinks below the tar.

Aside from being pretty effective at capturing animals, tar is very good at preserving bones as it solidifies over the years into black rock. True, the bones stink of petroleum and have a nasty yellow look to them, but they are otherwise in perfect condition.* The only catch is that sorting out the fossils that result from the scavenger/predator/prey chaos is something of a paleontological nightmare. Bones from the original thirsty animal get mingled with those of the predators, and bones from predators get mingled with those of scavengers. It can take thousands of hours to pick apart the fossils left behind and work out how many animals there were and what sorts of relationships they had with one another.

With tar pits and their processes firmly in mind, consider the following story. A thirsty goat wanders down to a quiet lake. This watering hole does not appear to be staked out by predators as so many other watering holes are, and the goat steps into the shallows for a quick drink. Suddenly, it finds that it cannot pull its legs back out of the lake. Indeed, the reason the watering hole is not staked out by predators is because it is not a watering hole at all, but a tar pit. The goat gets thoroughly stuck as it struggles to get out, and its screams attract a lion. The lion pounces on the goat and gets stuck as well. Both the lion and the goat die, but before they sink beneath the tar, vultures come in to feed. These vultures, because of their light weight, do not get stuck, but their presence attracts a bird-hunting viper that slithers over the edges of the tar pit without any trouble but gets stuck in the recently churned-up tar where the lion and goat died. All three, the goat, the lion, and the snake, slowly slide into the

*The University of California, Berkeley, has a very large collection of these sorts of fossils excavated from tar pits in Southern California but is unable to store them with the rest of their fossils in the university's paleontology museum because of the stench and the dangerous nature of the petroleum fumes. The university now keeps them in the bell tower at the center of the campus where they can "de-gas" in peace.

tar together, get preserved, and their bones are ultimately found in stinky blackened rock by people who can only wonder at what sort of creature would have left such a skeleton behind.

And Chimera was hardly alone. If a horse went down to a tar pit for a bit of water, got stuck, died, and was subsequently fed upon by a vulture that also got stuck in the tar, that would provide an explanation for the legendary Pegasus. Some art even shows Chimera battling with Pegasus. Was this linked to a find of fossils that people could barely make sense of?

Chimera Versus Pegasus, attributed to Heidelberg painter. Greek cup, c. 560–550 BC. Louvre Museum, Paris. Art Resource, NY.

The Sphinx could have also had her origin here.* With the body of a lion, the head of a human, and the wings of an eagle, it is reasonable to propose that a human or an ancestor of modern humans like

*According to Hesiod's *Theogony*, the Sphinx, Chimera, and Cerberus were siblings. One has to wonder if their sibling status was invented to explain them all being found in a similar location or in a similar fossilized state.

Homo erectus went down to a water-covered tar pit for a drink, got stuck, was attacked by a lion, and then the dying lion was feasted upon by an eagle. All got trapped and were preserved together. This, of course, raises the question of whether humans or their ancestors were stupid enough to go down to a tar pit thinking it was a safe place to get water or collect plants. When visiting tar pits today, even disguised ones, the scent of petroleum is always thick in the air and the area seems "not quite right." But perhaps it took years of trial and error for the lesson to take hold, since fossils of humans and their ancestors do, very occasionally, turn up in tar pit fossil sites.

Among the ancient monsters that tar pits best explain is Scylla. First extensively mentioned in Homer's *Odyssey,* Scylla lived on one side of a narrow channel opposite a monstrous whirlpool known as Charybdis that ate ships for breakfast (more on Charybdis in "The Mysterious Fathoms"). Odysseus had to sail through this channel and was forced to decide whether he would sail within reach of Charybdis or within reach of Scylla.

Chimera's complex anatomy pales in comparison to Scylla's. She had four eyes, six heads that each had three rows of sharp teeth, twelve legs, a feline tail, and numerous wolf or dog heads attached to her waist. Intriguingly, while Scylla is well known from later Greek art and stories to have had the upper body of a female human, her earliest depiction, on an Etruscan cylindrical ivory box covered in reliefs from 600 BC, is not human at all. The monster has multiple wolf or dog heads sticking out from several serpentine necks. Indeed, if there is a monster that stands as evidence that the ancients were looking at fossils of multiple animal skeletons jumbled together, it is Scylla.

However, this requires tar pits, and Greece (and the rest of Europe) doesn't have any. In the heart of Los Angeles, thousands of predator fossils have been dug up. Indeed, a museum at the site has a wall of wolf skulls beautifully arranged to drive home the point that this site was a very effective predator trap. Yet the La Brea Tar Pits would have been entirely unknown to the Greeks, who knew nothing of the New World.

The reason California has a lot of tar pits is because tar and oil tend to go hand in hand. Where there is oil, tar is often not far off. Not surprisingly, tar deposits are found in Russia, a major modern oil producer, and in the Middle East, which also has lots of oil. So it might not have been the Greeks in Greece who initially conjured up Chimera, but rather, Greeks living in colonies along the eastern coast of the Black Sea (now Russia) or along the eastern and southern coast of the Mediterranean Sea, where some oil reserves are located not far from the Syrian coast. Trade routes certainly ran from these colonies to Greece, and the tales of strange fossils may have spread along these routes.

It is possible that the idea behind a monster like Chimera goes even farther back, to the Kingdom of Israel between 1010 and 930 BC, when Israelites ruled the land near the inland Syrian oil fields of Tadmur. These people, or even an earlier civilization, may have seen mixed and mangled bones and come up with the ideas that ultimately coalesced into Chimera. Regardless of where and when the monster was first imagined, it is not hard to envision the discoverers of a tar pit fossil spreading word of a terrible creature with a goat's body, a lion's head, and a snake for a tail.

The possibility of Chimera having arisen in the eastern Mediterranean is further supported by Homer, who specifically states that the monster dwelt in Lycia, a region distinctly outside ancient Greece. It would be wonderful—not to mention a tremendous boon to the tar pit argument—if Lycia contained tar pits full of fossilized predators, but Lycia does not have any oil or tar pits. Lycia does have interesting geologic features that are worth noting, though. Both Homer and Hesiod say that Chimera had the ability to breathe fire. Homer specifically describes it as "snorting out the breath of the terrible flame." Intriguingly, Lycia is one of a few places where natural gas slowly leaks out of Earth's crust to the surface. Today, people light this gas and, because there is a limitless supply seeping out from below, the flame never goes out. Before human mastery of fire, people would have been unable to light this gas, but a single wildfire in the area resulting from a lightning strike, could have changed that in a hurry. Once lit in such a way, the flames would have kept on burning, leading those who discovered them to wonder what could cause fire, with seemingly no source, to burn for such a long time.

For traders and travelers coming from communities along the eastern coast of the Mediterranean Sea, Lycia was on the way to Greece. Grains, copper, iron, dates, and wine all migrated along Greek and Phoenician trade routes that started where Israel and Syria are today and hugged the southern Turkish coastline. Some traders would have undoubtedly stopped for rest and resupply in Lycia. Did those who saw strange mixtures of bones in stinking blackened petroleum-filled rocks in the Middle East make a connection with the stench of burning petroleum gas at Lycia and bring the two elements together in the form of the fire-breathing Chimera? The possibility is a tantalizing one. Certainly the initial discovery of strange-looking bones and flames leaping out of the earth all on their own must have scared the hell out of people. The ingredients were there for the birth of a monster, but there is even more to fear in the creature when deeply seated psychological tendencies are taken into consideration.

Monstrous mélange

At its core, Chimera is an aberration, a creature that deviates from normal biology in an extreme way. Make no mistake, it likely arose from fears of the unexplainable features that people were finding in the natural world, but it endured because these natural observations led to the imagining of something quite horrible. Chimera's weird blend of animals made it a vile sight to behold, and science is now suggesting that humans may be hardwired to react negatively to its alien body structure.

A topic of frequent discussion among researchers who study human mate selection and health is symmetry. A number of studies over the years have explored how people view others with perfectly symmetrical faces and faces that are somewhat asymmetrical. It turns out that people associate symmetrical faces with increased health and are more attracted to them. In contrast, asymmetrical features are seen negatively.

Growths that make one part of the face become deformed, mutations that cause people to develop six fingers on one hand, wounds, jagged scars, and malformed spines that cause one shoulder to become significantly higher or lower than the other are all, sadly, viewed with various levels of distaste. The evolutionary reasons for this are widely thought to be associated with fitness. Like all animals, humans are driven to reproduce with partners who will help them have lots of healthy children. If a potential partner has numerous scars and/or malformations, these could be associated with poor reflexes, a weak immune system that has allowed the person to suffer from considerable disease, past episodes of malnutrition, or bad genes. Thus, the evolutionary argument goes, we view these traits negatively because they hint that a potential partner carries genes that we don't want to mix with our own. This negative view of asymmetricality is not limited to attraction; it has spilled over into popular culture and become associated with overall "badness." This is why villains like the six-fingered man in *The Princess Bride* and the lion Scar in *The Lion King* are so often associated with these sorts of characteristics. In fact, their names *are* their deformity, because the negative essence of the trait represents them so completely.

Looking at Chimera is, in effect, not much different from looking at a human who has two heads, a scar running across one eye, or a missing limb, and this adds to the monster's fear factor.

Chimera itself does not feature much in modern books, films, or television, but many of the fears that it embodied are alive and well. When H. G. Wells wrote *The Island of Dr. Moreau* in 1896, he was looking out upon the dawn of a scientific era when surgery and veterinary science were beginning to suggest that biological tampering might make it possible to merge human and animal features. The scientist, Dr. Moreau, uses his knowledge of physiology to make animals more human, effectively creating creatures that are neither man nor beast. These beast folk strive to throw off their animal instincts and struggle to follow a strict code of laws in their primitive society.

They are forbidden from hunting, chasing, walking on all fours, and lapping up water with their tongues, and they continually repeat the mantra "Are we not men?" Yet as the story unfolds, animal instincts prove difficult for the beast folk to control, and the island slowly collapses into dangerous disorder with the question "Are we not men?" resoundingly answered with a no.

Wells makes a clear argument that tinkering with life by using surgery to try and make animals more human is something dangerous. Prendick, the shipwrecked protagonist, finds the beast folk horrific, and many of his encounters with them in the jungle interior of the island are the stuff of nightmares. If the idea of humanized animals brings to mind ancient monstrous creatures that merged human and animal features like the Sphinx and the spine-tailed Manticore, that should only further drive home the point that mixing human and animal traits creates creatures that have terrified people for millennia.

Yet in *The Island of Dr. Moreau* there is an intriguing contrast to the mixed monsters of ancient history. While Chimera, Cerberus, Manticore, Scylla, and others were monsters created by the gods, the beast folk of Dr. Moreau's island are entirely the result of a single man recklessly wielding science. Indeed, just as *King Kong* is clearly a monster movie with a monster that is not so easy to vilify because of human cruelty toward the giant ape, *The Island of Dr. Moreau* is a monster story where the monsters are really rather pitiful victims of the doctor's scientific work. That Dr. Moreau is a villain is readily apparent.* But where exactly the monster in the monster story resides is hard to say, since the two key elements found in monsters—a visage of horror and a willingness to harm others—are somewhat divided.

Unsurprisingly, *The Island of Dr. Moreau* has found its way onto the silver screen three times since it was first written, and in the latest

*Heroes don't usually say things like "Each time I dip a living creature in the bath of burning pain, I say, 'This time I will burn out all the animal; this time I will make a rational creature of my own!'"

(and not particularly good) adaptation, released in 1996, it embraced the rising use of genetics. Shifting from Wells's original suggestion that the doctor was surgically humanizing beasts, the recent version describes the doctor as using drug therapy and gene manipulation to accomplish the same result. Along the same lines is Rupert Wyatt's 2011 film, *Rise of the Planet of the Apes,* which links the testing of a newly developed Alzheimer's drug on primates with the development of human intelligence among the apes themselves. Like *The Island of Dr. Moreau, Rise of the Planet of the Apes* again raises the question of who or what the monster actually is. Humans certainly have their moments of villainy in the story, but the apes, with their disturbingly human facial characteristics and ability to so effectively slink through the shadows once they escape from captivity, definitely are frightening. Are they the heroes? It certainly feels that way when a valiant gorilla sacrifices itself to save the leader ape, Caesar, from gunfire. Indeed, there is much here that is similar to *King Kong,* with sympathy building for the creatures that would typically be identified as monsters. That there is a lot of ambiguity is unquestionable and, based upon where modern science is headed, understandable.

The creatures of Wells's imagination are not as far from reality as they might seem. Numerous mice, rabbits, sheep, fish, and birds have already been genetically engineered to carry and express the genes of other animals, including the genes of humans. The methods that are used vary. Some techniques directly insert genetic material from one animal into the area where the genes of a developing egg cell are found. Some labs are engineering viruses to carry genetic material and inject it into the newly developing cells of an embryo. Perhaps the most widely known technique adds genes to stem cells, which have the ability to become other types of cells in the body. By altering stem cells in this way and then adding these altered stem cells to a developing embryo, the added genes become expressed as the embryo grows.

Such techniques have already made it possible for teams to create a mouse with the liver of a rat. This might not sound like a big deal, since mice and rats are closely related, but giving one species the ability to grow and live off of the organs of another is not that far from what Wells was writing about.* Indeed, the journal *Nature* published an editorial in 2011 titled "The Legacy of Doctor Moreau," arguing that even though the blending of animal and human characteristics will likely be viewed by modern audiences with the same level of horror as Victorian audiences greeted Wells's beast folk, such horror must be overcome for the sake of science and properly managed by a well-established framework of rules.

A mouse with a rat liver does not inspire horror among the public, but what about a rodent born with furless human skin? The skin is, in fact, just another organ and, genetically speaking, creating such a rodent in the lab is something researchers are on the verge of doing. This will likely be met with widespread revulsion,† but for the sake of finding treatments for life-threatening skin diseases, like skin cancer, should such revulsion be overcome? *Nature* certainly argues for this, with the caveat of careful government oversight. But what of a monkey being born with a human brain?

Based upon how far research has progressed in recent years, such a creature is now no longer outside the realm of possibilities. Although it will not be created in a lab tomorrow or next year, in the coming decades an animal of this sort may well become very real. But what would such a creature endure? We would have to apply all ethical regulations afforded to humans to a monkey with a human brain, but the mere possibility of such an organism being created and the terrible questions that such scientific work raises are understandably frightening. Could it ever learn to speak? Would it go insane? Might it resent its creators and plot revenge?

To help keep the nightmares at bay and maintain some level of ethical control, the Academy of Medical Sciences in London has set

---

*Remember, humans and chimpanzees are actually more closely related to one another than mice and rats.

†And fiery protests that will make all the battles fought over stem-cell research look like a picnic in the park.

out a number of rules intended to guide genetics work during the years ahead. Among other things, it clarifies the practices that should be considered reasonable during the introduction of human stem cells into animals that lead to the creation of "chimeric" embryos.

Chimera, still very much alive, still generating fear, and most certainly coming soon to a cinema near you (probably in the form of a vengeful monkey with a human brain).

3

It Came from the Earth—Minotaur, Medusa

"Snakes. Why'd it have to be snakes?"

—Indiana Jones, *Raiders of the Lost Ark*

While walking along the sun-baked hills of the Greek islands and staring out over the sparkling blue water, it is hard to think of anything other than paradise. There are no dark forests, no tar pits, and no fierce beasts, just the gentle sea breeze. It is hardly the sort of place where one would expect to find a monster, yet on the island of Crete, one of the earliest and fiercest of monsters came into existence, and not on the surface of the island but in the subterranean world below it.

Half bull and half man, the Minotaur lived deep in an underground labyrinth and captivated the ancient Greeks who frequently portrayed it in their art. Some drew it with the head of a man and the body of a bull, making it look kind of like a Centaur. Others presented it as a man with a bull's head. Many artistic depictions of the beast show it attacking and eating people.

The myth, as described by Apollodorus, tells of a man named Minos on the island of Crete who sought to become king: "Minos aspired to the throne, but was rebuffed. He claimed, however, that he had received the sovereignty from the gods and to prove it he said that whatever he prayed for would come about. So while sacrificing to Poseidon, he prayed for a bull* to appear from the depths of the sea, and promised to sacrifice it upon its appearance. And Poseidon did send up to him a splendid [white] bull."

Yet Minos proved greedy. After becoming king, he decided he liked the bull so much that he kept it as a pet rather than sacrifice it. This angered Poseidon, who "devised that Pasiphaë [Minos's wife] should develop a lust for it [the bull]. In her passion for the bull she took on as her accomplice a genius architect named Daidalos. . . . He built a wooden cow on wheels, . . . skinned a real cow, and sewed the contraption into the skin, and then, after placing Pasiphaë inside, set it in a meadow where the bull normally grazed." Minos's wife wound up getting pregnant by the bull.† Nine months later, when she gave birth, the result was the Minotaur.

Minos's wife nursed the monster during its early years until the beast started eating people in the household. Minos was understandably concerned, but since he had already wronged the gods by keeping the white bull as a pet, he was not eager to anger the gods again. The ever-helpful Oracle at Delphi confirmed that his instincts were correct and that he would indeed be in serious trouble if he plotted the death of the Minotaur. This left Minos in a bind. He could not kill the beast, but letting it roam the palace where it could eat everyone was out of the question. To solve his problem, Minos had Daidalos build a maze underneath the palace where the Minotaur could effectively be imprisoned.

---

*When wishes show up in the literature of other civilizations, the people who make them typically wish for riches, visions of the future, or eternal youth. Not the Greeks. No, Minos wished for a bull. Go figure.

†Don't ask.

Minotaur. Greek, Attic bilingual eye-cup. c. 515 BCE. Art Resource, NY.

With the Minotaur stuck in the labyrinth, one might think the story of the monster safely concluded, but the beast continued to cause trouble. The Greek poet Callimachus described it making "cruel bellowing" from its labyrinthian jail, and so to keep the monster calm, Minos arranged to have it fed foreigners on a regular basis. Dozens of hapless people met their fate in the maze, but eventually Minos made the mistake of sending the Athenian hero Theseus inside. Using a ball of string given to him by Minos's love-struck daughter Ariadne to leave a trail behind him, Theseus killed the Minotaur and escaped the maze.

The idea of a Minotaur is firmly in the realm of fantasy and myth. There are no animals alive today or found in the fossil record that combine the traits of humans and bulls in any way. Moreover, even if such a creature did exist, the biology would not work. Unlike, say, humans and bears, which have teeth and digestive systems that can manage both plants and meat, bulls are obligate plant eaters and cannot chew or process meat. Since there are no fossil mammals that suggest the merging of a bull and human skeleton, it is more likely that the concept for a beast that was half man and half bull stemmed from Greek interpretation of the culture found on Crete between 3000 and 1100 BC.

The people of ancient Crete, the Minoans, were ahead of their time. Women are depicted in paintings as having been leaders. The cities had intricate plumbing systems. Tools of war are almost entirely unseen in the archaeological record. Paintings of dolphins and bulls abound with art revealing Minoans engaging in games with bulls, grabbing them by the horns, and vaulting off of bulls' backs. They are somewhat reminiscent of Spanish bullfights except that no works of art have yet been found showing a Minoan attacking or slaying a bull. Archaeologists speculate that Minoans were engaging bulls for sport and that the art depicts a favorite activity of the culture.

By the time the Greeks emerged as a strong and healthy civilization, the Minoan world had effectively collapsed. The reasons for the collapse are debated. Some suggest that invaders with iron weapons overwhelmed them while others posit that a tsunami or severe earthquake wiped out the society. Regardless of the cause, it is possible that the mythmaking Greeks heard stories, passed along by word of mouth, of a people on Crete "who were one" with bulls. Some artwork showing humans and bulls grappling may have further inspired the idea of a half-man, half-bull monster.

But the Greeks were not the first to merge man and beast. Ancient paintings in the Chauvet cave of southern France, where some artwork on the walls is nearly thirty-two thousand years old, depict a creature that is clearly half woman and half bull. Whether this was a monster is a mystery, but we know these ancient humans were not bull leaping like the Minoans. What seems a more probable explanation for this drawing is that early humans saw the power of wild animals and believed drawings that mixed animal and human features imbued some strength of the wild upon them. The same ideas may have been present in Greece and played their part in the rise of the Minotaur.

Even so, these arguments at best only partially explain the story of the monster. Why go to the effort of inventing the labyrinth, and what about the "cruel bellowing" that Callimachus describes as having come from belowground?

If it truly existed, exactly where the labyrinth was built is a matter of intense debate among historians. Sir Arthur Evans, the first person to excavate the Minoan palace of Knossos on Crete, proposed that the extensive ruins of Knossos itself, which have numerous tunnels and passages carved into the ground, was where later

Greeks believed the labyrinth to be. Others argue that the labyrinth was associated with either the Skotino cave complex on the island, a short journey east of Knossos near the modern resort village of Gouves, or a series of tunnels about 45 miles south of Knossos at a site known today as Gortyn.

Regardless of the precise location of the labyrinth, there is no question that the Minotaur was specifically being placed underground and that its "cruel bellows" were coming from some subterranean location. To fully understand what this means, a bit of geology is required.

Shaken and stirred

Earthquakes have been common on Crete for more than a hundred thousand years. The reason is because Crete, and many of the other Greek islands, are sitting above some very active sections of Earth's crust.

Not all crust is created equal. Some crust forms the continents and some crust forms the ocean floor. At first glance they look the same, but they are chemically and functionally quite different. Continental crust is relatively light and buoyant, while oceanic crust is relatively dense and heavy.

In areas called subduction zones, like those around Japan, Indonesia, and Washington State in the United States, heavy oceanic crust moves toward the coasts and runs into buoyant continental crust. When this happens, the ocean crust's weight draws it under the continental crust and it begins a long journey deep into the earth. During this journey, the oceanic crust cracks, causing earthquakes, and experiences ever-increasing pressures and temperatures that cause it to melt. Because heat rises, some of this molten rock comes up toward the surface and eventually gets blasted out of volcanoes. These regions are almost always marked by a series of big volcanoes laid out in rows aligned with the subduction zone.

On Crete, the geology is different. South of the island there is an enormous plate of continental crust that makes up most of Africa, and attached to the north side of this plate is a bit of old ocean crust. Crete itself is sitting on a small plate of continental crust, known as the Aegean Plate, that makes up both the floor of the Mediterranean Sea and the islands in the area.*

Unlike other subduction zones, where the ocean crust is doing the moving and sliding under immobile continental crust, in the Mediterranean, the continental crust of the Aegean Plate is sliding southward onto the bit of oceanic crust sitting along the northern tip of the North African Plate. Moreover, it is moving at the very fast† rate of 1¼ inches (33 millimeters) a year while the North African Plate is moving northward at the sluggish rate of ⅕ of an inch (5 millimeters) a year.

With these differing speeds and movements, Crete ends up in an unusual situation. The oceanic crust attached to the North African Plate is not subducting in the way that ocean crust normally subducts in other parts of the world because there is not very much of it, it is not very heavy, and it is attached to the enormous chunk of continental crust making up North Africa. It is still melting and ultimately forming volcanoes far north of Crete, but rather than going down at a steep angle as oceanic crust usually does, it is staying stubbornly shallow and forcing the Aegean Plate upward. Indeed, Crete exists as an island specifically because the North African Plate is constantly pushing it up and out of the sea at a rate of ⅕ of an inch annually. While a rise of ⅕ of an inch doesn't sound like much, it is a lot for an island to grow in a year.

As for earthquakes, Crete's position is a miserable one. Because the ocean crust on the northern tip of the North African Plate is subducting (admittedly in an odd way), this causes earthquakes as

---

*Continental crust is defined by its chemistry, not by whether it is above or below water. It can have water on top of it; this just doesn't happen very often.

†This is, of course, relative. Aeronautical engineers get excited by planes that can break the sound barrier. In contrast, geologists get all excited when they find crust moving at a rate faster than your fingernails grow. An easier crowd to please.

the old crust cracks during its shallow descent. In places like Japan and Washington State, which have large earthquakes, this would be the end of the geologic story, but in Crete, there is more. The Aegean Plate's rapid movement up and over the North African Plate is causing tremendous tension to build up quickly and leading sections of both plates to crack more often than they would if one plate was just slowly descending below the other. This makes Crete much more prone to earthquakes than most other parts of the world.

Could regular earthquakes be the origin of the Minotaur myth? Such tectonic activity could have left ancient inhabitants of Crete searching for stories to explain what they were feeling beneath their feet. Certainly, the sound of an earthquake could be described as the "bellowing" of a beast somewhere underground. But it seems more likely that there must have been some unbelievably severe earthquakes that wreaked such terrible havoc that an explanation, like that of the Minotaur, was needed. And written accounts reveal that horrific earthquakes are a part of Crete's past.

On July 21, 365 AD, the Roman historian Ammianus Marcellinus, who was living in Alexandria, wrote:

> Slightly after daybreak, and heralded by a thick succession of fiercely shaken thunderbolts, the solidity of the whole earth was made to shake and shudder, and the sea was driven away, its waves were rolled back, and it disappeared, so that the abyss of the depths was uncovered and many-shaped varieties of sea-creatures were seen stuck in the slime; the great wastes of those valleys and mountains, which the very creation had dismissed beneath the vast whirlpools, at that moment, as it was given to be believed, looked up at the sun's rays. Many ships, then, were stranded as if on dry land, and people wandered at will about the paltry remains of the waters to collect fish and the like in their hands; then the roaring sea as if insulted by its repulse rises back in turn, and through the teeming shoals dashed itself violently on islands and extensive tracts of the mainland, and flattened innumerable build-

> ings in towns or wherever they were found. Thus in the raging conflict of the elements, the face of the earth was changed to reveal wondrous sights. For the mass of waters returning when least expected killed many thousands by drowning, and with the tides whipped up to a height as they rushed back, some ships, after the anger of the watery element had grown old, were seen to have sunk, and the bodies of people killed in shipwrecks lay there, faces up or down.

Marcellinus was witnessing a tsunami of tremendous proportions, and according to research studies of the area, it appears that the inhabitants of Crete were plagued by these disasters.

In 2007, a team of geologists led by Beth Shaw, then a researcher at the University of Cambridge, set out to determine what might have caused the ancient tsunami and whether there was still a threat of enormous waves to the region. The team's work brought them to Crete, where they started analyzing the carbon found inside corals.

Carbon is an element that appears naturally in the environment in two common forms, carbon 12 and carbon 14. Carbon 12 is stable and, as long as it is left alone, remains carbon 12 in perpetuity. Carbon 14 initially forms as the result of cosmic rays striking it and is a radioactive material, meaning that over time it loses energy by radiating away energized particles. During their lifetimes, all animals consume both carbon 12 and carbon 14 from their environment. After they die, the amount of carbon 12 in the body remains constant but the carbon 14 loses energy and slowly degrades into carbon 12. All samples of carbon 14 shed energy at precisely the same rate and, for this reason, it is possible to look at the carbon 14 found in fossils and work out very accurate dates of animal death by comparing the amount of carbon 14 in the fossil to the amount of carbon 14 in animals that are currently alive.

What surprised the Cambridge team was that all of the algal organisms and aquatic tube worms that had been growing along the western shore of Crete suddenly died in AD 365. When they looked closer, they noticed that the ground around the fossilized organisms had marks on it that looked like a bathtub ring around the island. They realized that tectonic activity had suddenly shoved the western side of the island up and out of the Mediterranean Sea. The team's analysis, reported in *Nature Geoscience* in 2008, proposed that tectonic activity pushed the western portion of Crete nearly 32 feet (10 meters) out of the water in a single powerful tectonic event. This sudden uplift quickly dried out all of the algae and worms that had been sitting happily on the seafloor and killed them, thus explaining why they all have exactly the same carbon 14 value.

Lifting that much rock out of the water would have caused a tsunami of epic proportions,* similar to what Marcellinus described, as well as been connected to a very powerful earthquake. Yet 365 was several thousand years after the days of the Minoans and was probably just an isolated tragedy, right?

Not so. Work on the geology of the Mediterranean, conducted by Anja Scheffers at Southern Cross University in Australia and reported in *Earth and Planetary Science Letters* in 2008, revealed boulders along the Greek coast with shelled marine animals attached to them. Scheffers and her team realized that, because of the animals attached to them, these boulders had to have once rested beneath the waves. Yet the boulders themselves were not sitting among marine sediments. The only logical explanation was that giant waves had thrown the boulders out of the sea and onto dry land while animals were still hanging on.

When the team analyzed the carbon 14 of the fossil animals on boulders, they found that some of the boulders had been thrown out of the water in 365. This was not surprising. However, they also found evidence of tsunamis having formed in and around 6000, 1000, and 200 BC, suggesting that really big tsunamis, and the powerful earthquakes responsible for creating them, have been common phenomena along the Greek coast for the past ten thousand years.

*We're talking a real wrath-of-God-type event, the sort of thing that would turn even an ancient Richard Dawkins into a believer.

Scheffers does not provide any evidence of powerful tsunamis between 3000 and 1100 BC, when the Minoan civilization was flourishing, but there is no need. Earthquakes half as strong as those that can send boulders flying out of the ocean terrify people living today who understand the geologic forces behind them. To the poor souls living on ancient Crete, violent shaking of the ground had to have been downright petrifying. During these episodes, buildings would have quickly crumbled, with collapsing pieces of stone and wood snapping human bones as if they were twigs. Without even the most rudimentary geology available to turn to for an explanation, there was no denying that these destructive events were the result of the Minotaur and the deafening roar was its angry bellowing as it fumed in misery deep within its shadowy labyrinth.* But not all people of the world have come to associate earthquakes with monsters. In some places, earthquakes are viewed as indications of good deities being present.

Beyond the labyrinth

On the Hawaiian islands, where people have long been relentlessly exposed to strong earthquakes, natives worshipped the goddess Pele. Described in myths as a sensitive woman of beauty with a fierce temper, she represented both the destructive fury of volcanic eruptions and the mesmerizing enchantment of Hawaiian dance. Sure, touching her could set fire to a mortal's skin, but she was also sensitive, loving, and emotional. Thus Pele, the fair and fiery, is a far cry from the Minotaur.

Hawaii is nowhere near a plate boundary. Indeed, the volcanic eruptions and earthquakes that are so common there are not related

---

*There is the added fact that the bull, which started the entire Minotaur mess, was given to King Minos by Poseidon, the god of both the ocean and earthquakes. Mere coincidence?

to plate subduction. Instead, Hawaii is the result of what geologists call a hot spot, which is a section of Earth's mantle that is superheated and burns through to the crust, making it thin. Sometimes groundwater in this crust on continents becomes so hot that it bubbles up to form geysers and thermal pools like those seen at the hot spot that is Yellowstone National Park. Where Earth's crust is the seafloor, the eruptions form islands of volcanic rock, as is the case with the hot spot responsible for Hawaii.

The islands that ancient Hawaiians lived on were all formed from lava released by volcanic eruptions. Islanders would have witnessed many eruptions that blanketed the land in fiery red magma. This magma, which would have been destructive and lethal at first, would then solidify and, in areas near the edges of the islands, pour out into the ocean, creating new land. In just a single human life span the magma would cool, soften with exposure to the elements, and transform into fertile soil that would rapidly become covered in verdant plants.

With volcanic eruptions so obviously tethered to the formation of their own island and the creation of life-granting soil, it hardly seems surprising that tectonics took the form of the loving, creative, and fierce Pele rather than just a destructive force to be feared. The Minoans, on the other hand, would not have seen any beneficial effects from the many earthquakes they experienced, in spite of the fact that these earthquakes were the cause of their island's very existence. Even in the case of a major uplift event, a Cretan coastal shelf suddenly raised up and out of the water would not have become covered in lush vegetation because Crete is not a place where dense plant life readily sprouts.

Today, while Minotaurs are, thankfully, absent, portraying the Earth as a monster in the media continues to prove popular among those who like a good scare. The reason, quite possibly, is because the unknowns of geology are conceptually similar to the dark shadows of the world's jungles.

The films *Volcano* and *Dante's Peak* portrayed heroic individuals racing against time to save the great masses from the imminent eruption of a volcano. *Dante's Peak* has some credible science behind it. The story features a town in the mountains that sits next to a volcano that has not erupted for seven thousand years. Suddenly, for no apparent reason, the volcano becomes active. First, visitors to hot springs get boiled alive as the springs leap to lethal temperatures. Then, as the volcano really gets going, the acidity of water in the surrounding area spikes from all of the sulfuric gases released by nearby volcanic vents. Eventually, there is a catastrophic eruption and much Hollywood-friendly action ensues. The timeline of events in *Dante's Peak* is faster than would likely be seen in the real world, but the unpredictability of the volcano and the sorts of threats an eruption might really pose are reasonable enough.

In contrast, the volcano in *Volcano* appears underneath the La Brea Tar Pits in Los Angeles shortly after an unexpected earthquake. Superheated steam kills people in underground tunnels, magma begins to ooze up from the ground, and fires threaten to destroy the city. Rather than run away, the protagonists arrange to "redirect" the lava out of the city and into the ocean. More terrible ideas for a film about a volcanic disaster would be challenging to come up with.

Volcanos simply do not form anywhere near as quickly as they are shown to do in the film and, as much as some may like there to be a volcano simmering below Los Angeles, no such thing exists. Moreover, tar pits do not have any known link to volcanic activity and magma, which is hot enough to melt stone, cannot be redirected around a city using lane dividers on a freeway.

Even so, as stupid as *Volcano* is, it and *Dante's Peak* play off of real unknowns associated with when and why earthquakes and volcanic eruptions happen. Among geologists who have studied regions like Yellowstone and Hawaii extensively, there is simply no consensus on what causes these volcanic hot spots to exist. Some argue that they are the result of unusual, localized chemistries in the mantle that cause it to become superheated. But nobody is sure. And with the creation of a monster, what better place to start than with an aspect of the natural world that is barely understood? Just like the Minotaur, the volcanic eruptions in the films are beasts built from observations of geology that nobody can make sense of.

Will we see more volcanic eruptions portrayed as monsters? It is difficult to know. The Minotaur came into being in a region where earthquakes were common and explanations were few and far between. *Volcano* and *Dante's Peak* were both produced in 1997, which, in the movie-making world where production takes ages, was a mere blink of an eye after the 1991 Mount Pinatubo eruption in Indonesia, which killed roughly eight hundred people, left a hundred thousand homeless, and was presented in live video on news channels around the world. While it is impossible to leap into the minds of the producers and screenwriters and determine what exactly inspired them to develop these films, a safe bet is that discussions of volcanic eruptions in the news were involved.

We will no doubt see many more monsters take the form of tsunamis, earthquakes, and volcanic eruptions in the years ahead, yet it would be foolish to believe that tectonic movements are required for Earth to inspire monsters. Many processes far more subtle than those associated with earthquakes have created fears that have left sturdy souls trembling in terror.

Secret in the stone

The Minotaur is not the only ancient monster to have come from fears associated with the earth. In the earliest stories from the *Theogony,* as told by the Greek poet Hesiod around 700 BC, Medusa is one of three daughters, known as the Gorgons, born to the sea gods Phorcys and Ceto. Further details in this early tale are lacking, but fortunately ancient art offers more, with Medusa often being portrayed with a horrible visage, snakes in her hair, fangs in her mouth, and, sometimes, wings on her back.

The playwright Euripides, who lived between 480 and 406 BC, mentions in his tragedy *Orestes* that merely looking at Medusa and her sisters could turn people to stone. Specifically, Orestes asks the Phrygian slave (or eunuch; translations vary), "Are you afraid of being turned to a stone, as if you had seen a Gorgon?" Euripides expected his audience to know that the Gorgons could petrify, and the allusion would have been clear. Further to this, a few hundred years later, a text that has often been attributed to Apollodorus states that Medusa and her sisters "turned to stone such as beheld them" and that "when her head had been cut off, the winged horse Pegasus sprang out of the Gorgon, and also Chrysaor [a warrior (or in some versions a flying monster) carrying a divine golden sword]."

Perseus Battling Medusa. Pitcher. ©The Trustees of the British Museum. All rights reserved.

But it is the Roman poet Ovid, born 43 BC, who goes further and presents an elaborate story describing how Medusa became a monster. Perseus, who ultimately beheads Medusa, explains: "Her beauty was far-famed, the jealous hope of many a suitor, and of all her charms her hair was loveliest; so I was told by one who claimed to have seen her. She, it's said, was violated in Athena's shrine by Ocean's Lord [Poseidon]. Zeus's daughter [Athena] turned away and covered with her shield her virgin's eyes, and then for fitting punishment transformed the Gorgon's lovely hair to loathsome snakes."

This is a major change. Rather than being born a monster, Medusa is transformed into one by Athena as punishment for being raped by Poseidon.

There is no question that the early and late stories of Medusa differ significantly, but with the exception of Hesiod, who describes almost nothing at all about Medusa, one element is consistent: She always has a gaze that can petrify flesh. It is the key trait that gives her great power and makes her an object of both terror and fascination. Where did this deadly ability come from?

Some historians argue that the text attributed to Apollodorus, with its talk of Chrysaor with his golden sword and the flying horse Pegasus, is actually an ancient record of a volcanic eruption. The golden sword, they suggest, is a bright orange burst of lava and the flying horse is a chunk of rock or burst of gas being thrown up into the air. Some even go so far as to suggest that the snake attributes associated with Medusa stem from snakelike rivulets of magma coming down the side of a volcano. To a certain extent, ash and lava can cover things and make them look as if they have been turned to stone. Certainly, in the Roman town of Pompeii, which was smothered by ash from Mount Vesuvius in AD 79, humans were killed by toxic gases and covered by ash, and the details of their bodies were preserved by the solidifying ash as their bodies rotted away. There may be some truth to all of this, but fossils provide an alluring alternative explanation.

Anyone with a keen eye who walks along a streambed or next to a cliff for any significant distance stands a chance of stumbling upon a bit of fossilized material. This can take the form of something as mundane as a fossilized shell or be as impressive as a dinosaur skeleton or a human skull. As different as these fossils appear, they all have something in common: They have been transformed from their original biological materials into stone.

Today this process is well understood. When animals die, their bodies fall to the ground and, under most circumstances, other animals come and consume the remains. This starts with the soft bits of the dead animal being ripped away by predators like lions, and generalist scavengers like most vultures. Then more specialized scavengers like hyenas and bearded vultures, which are adept at cracking open bones, break apart the skeleton to suck out nutritious marrow. If the remains are not entirely destroyed by all of this activity, bacterial rot, wind, and rain will usually finish them off in a matter of weeks or months.

However, in some environments, instead of being consumed or weathered, a body that falls onto the shore of a river, next to a lake, or adjacent to a sand dune can become covered by sediment that is either flowing by in the water or being blown about by the wind. Once buried, the remains are effectively locked away from the surface world.

This sort of burial scenario is common for animals that die by the mouth of a stream. Yet streams do not survive forever. Over time, a meandering stream can become a raging river or it can dry up completely. Such changes can happen in just a few thousand years or over the course of millions of years. Regardless of the time scale, a corpse buried by stream sediment can ultimately be exposed if the stream changes course and it begins eroding away the rock where the remains are located.

Throughout the journey from burial to eventual exposure, corpses are transformed. Flesh rots away. Bones, which are porous, are exposed to water that is traveling underground through sediment. Because groundwater is slightly acidic, it dissolves many of the materials it comes into contact with and carries these dissolved materials away. This water percolates through the many tiny holes in bones, and as it does so, the minerals in the water usually cause a buildup of silica, the primary component in sand, and eventually turn the bones to stone.* Remarkably, in spite of the radical chemical change involved in transforming bone to stone, the fossil that is formed from the process frequently looks very similar to the original specimen. Even subtle details, like partially healed injuries sustained by an animal during its lifetime, can sometimes be seen after fossilization.

For geologists, fossilized remains offer a glimpse of life from long ago. In contrast, ancient humans finding such things must have been perplexed. The more rationally inclined, like the sixth-century BC philosopher Xenophanes, speculated that the presence of fossilized shells high up on dry land meant that the sea must have once been a lot higher, but there was no understanding of fossilization at the time. During the late 1400s, Leonardo da Vinci went further by arguing that fossil shells he had collected were organic in origin. But it would not be until the 1600s, when Robert Hooke looked into his microscope at both petrified wood and normal wood and saw that they were remarkably similar in structure, that anyone would conclude there was a natural petrification process at work.

With fossilization being a total mystery for ancient humans, it seems likely that they tried to explain their findings by concluding there was a monster capable of turning things to stone. Yet even this explanation of Medusa's petrifying gaze raises questions. How did people make the connection between fossilization and gazing at a horrific face? Does fear have any psychological connection to stone?

---

*In some parts of the world, the mineral pyrite, more commonly known as fool's gold, can accumulate inside bone and transform it into a glittery replica of its original form. One has to wonder if the discovery of such transformed bones inspired the story of Midas, the mythical king whose touch could transmute objects into gold.

Frozen with fear

The expression "I was scared stiff" is so common in everyday language that few realize anyone can be harmed by fear. It is rare, but if scared badly enough, someone can actually suffer physical injuries. Medically, the effect is a form of shock.

But not the kind of shock that comes up in casual conversation. Being "shocked" to find out that a friend is cheating on a partner or that a loved one has unexpectedly died is not something that kills.

The type of shock that causes death has a single cause: not enough blood traveling to the brain and other organs. Emergency room personnel most commonly see hypovolemic shock, where massive blood loss means there's not enough blood left in the body to be pumped by the heart to the brain. The same sort of situation occurs with cardiogenic shock, where a heart attack causes so much damage to the heart that it cannot pump enough blood. There is a profound drop in blood pressure, resulting in inadequate blood flow to all the organs, including the brain.

Then there's neurogenic shock. Usually caused by physical trauma to the spine or skull, it disrupts neural signals sent to blood vessels by the brain. This is serious because these neural signals routinely tell the blood vessels to narrow or widen in order to regulate blood pressure. With the signals disrupted, the blood vessels relax and naturally widen. This causes blood pressure to rapidly drop because there is not enough blood to fill the (now wider) blood vessel system. With blood pressure too low, blood transport to the brain gets reduced or cut off entirely, unconsciousness follows, and shock sets in. Remarkably, giving somebody a sudden scare can also disrupt the neural signals to blood vessels and lead to a similar effect, which is commonly called psychogenic shock. This is what causes some people to faint at the sight of blood or upon seeing other psychologically distressing things.

Although the causes of these various forms of shock are different, they all result in the brain being denied the oxygen and glucose it needs to survive. When this happens, the brain starts having trouble functioning. The symptoms are often the same: Patients become disoriented, nauseated, and dizzy while their limbs become cold, tingly, and stiff as a result of reduced blood circulation. In a sense, the tingly paralysis that people often feel in their legs after sitting for too long is kind of like a localized, nonlethal taste of what shock patients feel in all their limbs. Moreover, a lack of blood flow to the brain can trigger seizures that sometimes lead to a state of boardlike stiffness in the body.

Human bodies worked the same way thousands of years ago and the ancient Greeks would have experienced the identical symptoms of psychogenic shock. With this in mind, it is tantalizing to consider the possibility that the concept of Medusa's petrifying gaze stems from someone who felt the icy grip of this shock upon seeing something frightening or worse, had a seizure from the experience, and entered a state of severe rigidity. But what of the strange notion that her hair was made of snakes? How did such a trait find its way into the myth of the monster, and what fears was this feature representing?

Sensible serpents

There are no living organisms that have heads covered in writhing snakes and, to date, there are no fossils indicating that such organisms ever existed. Why select snakes? Medusa could have had a head covered in tendrils like those on jellyfish or even tentacles like the limbs of a squid or an octopus. Yet the myths about her are consistent and specific: She always has hair of writhing snakes. This is not an accident.

Snakes kill people. Whether they are anacondas that squeeze the life out of humans stupid enough to swim in their jungle waters or cobras that inject venom with their fangs, snakes are a major threat to human life in many parts of the world today and there is evidence that this has always been the case. A 1978 study led by James Larrick at Duke University analyzed interactions between venomous snakes and the Waorani of Ecuador's Amazon rain forest. He found that 45 percent of the nearly six hundred individuals in the native population had experienced at least one snakebite and that about half of the people had been bitten more than once. As for the snakebite mortality rate, data collected on deaths over six generations indicated it had long hovered around 4 percent, roughly twice the 2.1 percent mortality rate that traffic accidents inflict on the developed world. True, with the invention of antivenins and emergency rooms, snakebites are now becoming less of a threat than they once were, but they were a terrible danger until very recently.

It is difficult to determine when highly venomous snakes and our primate ancestors first started mingling, since venom does not fossilize as bone does. However, venom injection mechanisms, like the hollow hypodermic needle teeth in venomous snake skulls, do sometimes get preserved and can hint at venom having once been present. Some fossil reptiles with hollow fangs have been dug up in Triassic sediments dating back to around two hundred million years ago, long before primates evolved. What these hollow fangs were used for cannot be determined for certain, but the paleontology team that made the find, led by Jonathan Mitchell at the University of Chicago, theorized in 2010 in the journal *Naturwissenschaften* that these ancient fangs were probably used for injecting venom. This suggests that venomous reptiles have been around since the days when mammals were only just beginning to evolve. Yet Mitchell's toothy find, named *Uatchitodon,* is more similar to a lizard than to a snake and is unlikely to have been on the same evolutionary path as modern snakes.

Other work, led by Bryan Fry at the University of Melbourne and published in *Nature* in 2006, analyzed large numbers of reptiles and studied the genes associated with their mouth secretions. While doing this, Fry and his colleagues noticed that some lizards had proteins in their mouths that were very similar to those associated with venom in rattlesnakes. This hinted that the common ancestor of lizards and snakes, which lived some one hundred million years ago, may well have already been carrying some mild toxins in its saliva and that these animals have long presented a threat to primates.*

Venomous snakes inhabited many of the landscapes where our ancestors evolved, and their venoms likely had the same effects then that they have on humans today.† Such realities are important to recognize, because with two hundred thousand years of cohabitation, there has been a lot of time for evolution to shape the way the primate mind responds to snakes.

In modern society, where warning signs can be posted in dangerous areas and word of mouth can advise people to be wary of well-known threats, the idea of evolving an intrinsic fear is not easy to grasp. With communication so readily available, we do not need to evolve a fear of things like speeding automobiles and electrocution, since there are plenty of people to tell us it is bad to be hit by a car or shocked by an electrical cable. However, for most living things, warnings are not available, and this would have been the case for our ancestors.

Imagine a male primate ancestor keen to have a romantic evening with an attractive female. Suppose, eager to impress, the male goes hunting for a rabbit that he can feed to his potential mate. If, while searching for the rabbit, he does not see a coiled viper hidden in the underbrush, his chances of stepping on it are increased. If he does step on it, he is likely to be bitten, and if he dies from the bite, he will not have sex, will not father as many children as males who avoid vipers, and fewer of his genes will be passed along to the next generation.

Now consider a male who happens to spot the coiled snake, backs away, catches the rabbit (or not), has the romantic evening,

---

*You didn't think it was just happenstance that the first antagonist in the Bible was the serpent in the Garden of Eden, did you?

†Snake venom varies a lot. Some snakes, like most of the vipers that inhabit Europe, deliver a bite that, with prompt first aid and basic medical care, causes only local tissue damage and leaves patients feeling numb and sick for a while. Other snakes, like the tiger snakes that inhabit the island of Tasmania, have venoms that kill within hours if antivenin is not administered quickly. Both would have probably proved lethal in an age when any serious weakness would have left humans vulnerable to predation.

and fathers numerous offspring. If the reason for the male noticing the viper had anything to do with genetics, which is possible since genes do code for things like the ability to detect color and motion, then his genes are going to be carried on to his young, who will also be likely to spot hidden snakes.*

What all of this means is that there was probably strong evolutionary pressure for our ancestors to be able to notice snakes as immediately dangerous so they could avoid them. And psychological studies seem to support this.

In a 2001 study conducted by Arne Öhman, a psychologist at the Karolinska Institute in Stockholm, participants were presented with both benign images of flowers and toadstools and threatening images of snakes and spiders. These images were presented nine at a time in a slide show format, and the participants were given a little keypad for each hand. One keypad was to be used if all the images seemed the same, the other was to be used if contrasting images were seen, like benign images standing out from threatening images or threatening images mingled among benign ones. The researchers carefully measured the time between the moment an image was shown and the moment the appropriate button was pushed. Öhman reported in the *Journal of Experimental Psychology* that students responded much more quickly when snake and spider images appeared than they did when presented with merely benign images. He concluded that modern humans have an innate ability to notice dangerous animals quickly. Moreover, he argued that this innate rapid detection ability came about because evolution has selected for humans who

*Color blindness is a genetic condition that commonly disrupts the ability to differentiate between the colors red and green. Intriguingly, research led by the primatologist Amanda Melin at the University of Calgary suggests that people who are red-green color-blind are particularly adept at spotting people who are wearing camouflage in forested settings. She and her team argued in the journal *Animal Behaviour* that this "disorder" may have once been beneficial to ancient humans who were hunting prey that was camouflaged. One has to wonder if it might also have played a role in helping them notice cryptically colored snakes on the forest floor.

are able to spot threatening animals and thus not become ill or die before they can reproduce a lot. And he was not alone.

Numerous studies show that many primate species have a considerable fear of snakes. "The west coast of Tanzania has lots of really dangerous snakes. Cobras, black mambas, green mambas are everywhere," explains primatologist Elizabeth Lonsdorf, director of the Lester E. Fisher Center for the Study and Conservation of Apes at Lincoln Park Zoo in Chicago. "People always ask me, 'Don't you ever worry about stepping on them?' and I always say no, because I follow the chimpanzees and they just seem to know to avoid them. When the chimpanzees see snakes, they climb a tree and make a very specific vocalization that sounds like a questioning 'hoo' while staring intently in the direction of the snake. All of us in the field learn the call real fast."

So why are snakes in Medusa's hair? Because snakes are immediately recognized as a potential threat and generate fear that naturally weaves its way into monster mythology.

Medusa modernized

Whether the concept of Medusa was scary to ancient humans is not much of a question. If people believed that fossils were the result of her actions and if the feelings associated with psychogenic shock were related to what they believed becoming petrified must feel like, then she would have seemed like a real threat.

During the long journey from the days of the ancient Greeks to modern times, Medusa is never forgotten. She is often depicted in art, with particularly famous presentations of her hideousness by Michelangelo Caravaggio, Peter Paul Rubens, Benvenuto Cellini, and even Pablo Picasso. But was she frightening to these artists?

Cellini's sculpture shows the hero Perseus, who slew Medusa, holding her severed head in victory. She is not alive and threatening but decapitated. In Rubens's painting her snakes are hissing frantically as the blood pours from her neck, and Caravaggio's Medusa has

Medusa, by Caravaggio. Oil on wood covered with canvas, 1570–1610. Uffizi Gallery, Florence. Scala/Ministero per i Beni e le Attività Culturali/Art Resource, NY.

a perfectly human face, admittedly ringed with snakes, that looks distressed. Her expression almost stirs a sense of pity. The decapitation is gross, but it isn't really scary or threatening. A painting could easily have been made showing Medusa about to attack Perseus or fighting with him in the shadows, ready to gaze into his eyes, but such works were never made. This is not to say that ancient art never showed her beheaded. Many works did present Perseus victorious, but they were balanced out by art showing Medusa and her sisters alive and dangerous. That from the Renaissance onward, Medusa is always shown as beheaded, beaten, and dying hints that while she continued to be a fascinating subject for artists to paint and sculpt, she was no longer a widely feared monster. In contrast, Medusa's decapitated head remained an object of fascination and, arguably, fear.

Late in the mythology surrounding Medusa, Perseus went on to marry the princess Andromeda, and during their ceremony a jealous suitor named Phineus* led an attack to claim Andromeda for himself. Phineus was a tough opponent and the battle was fierce, but it was never one that he was going to win. Aside from the fact that Perseus

*Who also happened to be Andromeda's uncle.

was part god, he was carrying around Medusa's severed head in a silk bag.* As Phineus lunged at the couple with his sword, Perseus shoved Medusa's head in his face, turning him and his minions to stone. Many artists from the Renaissance and later periods painted this scene, including Jean-Marc Nattier, Sebastiano Ricci, and Luca Giordano, and most made the transformation of Phineus's flesh from a healthy pink to a lifeless stone gray a focus of their work. In Ricci's art, men hold up shields to reflect the horrific sight, some lose their balance, and many are frozen in place as they attack. Similarly, in Nattier's painting, men scramble to hide their eyes from the petrifying gaze of the dead monster. There is an obvious sense of desperation that hints at something truly scary. Giordano shows Perseus wisely looking away as Phineus and his henchmen, with weapons drawn, are turned to stone.

Perseus with Minerva Showing the Head of Medusa, by Jean-Marc Nattier. 1729. Musée des Beaux Arts, Tours, France. Bridgeman-Giraudon/Art Resource, NY.

*Nuptial gift?

Medusa is never alive in any of these paintings—Perseus has only her chopped-off head—but the danger of a deadly gaze is always present. Were these artworks merely portraying a fear of the paralysis generated by extreme terror? Possibly, but there is another aspect of Medusa that could be playing a role. Medusa is distinctly female. From the very beginning she is always described as one of three sisters, and in later works, like Ovid's *Metamorphosis*, her femininity is accentuated by her description as a woman with beautiful hair who is even capable of attracting the attention of the god Poseidon. One has to wonder if these relatively late presentations of Medusa's deadly gaze are playing upon fears that men had of being overpowered by the gaze of a dangerous woman. To a certain extent these fears are still very much present in society (more on this later in "The Created").

Today Medusa is by no means gone. She was featured in both the 1981 and 2009 versions of *The Clash of the Titans* and in the (rather preposterous) 2010 film *Percy Jackson and the Lightning Thief*. This is a remarkable presence for such an ancient monster, but whether she still generates much of a fear factor is not entirely clear.

In Chris Columbus's *Percy Jackson and the Lightning Thief*, Uma Thurman portrays Medusa. The iconic writhing snakes for hair have, not very believably, been attached to her head, but she is otherwise shown as attractive. She wields no weapon whatsoever, has no claws or fangs, parades around a greenhouse in sunglasses, and depends upon words of manipulation and an attractive body to get people to open their eyes and look at her.

Like the Sirens of Greek mythology, which are at least partially scary because they make sensible people do life-threatening things (like sailing their ships into rocks), in this film, Medusa is a seductress whose threats are not immediately apparent. Her hideousness is absent and her petrification ability and snakes are so blatantly unrealistic that there is little to frighten anyone over the age of five.

Both Desmond Davis's 1981 *Clash of the Titans* and Louis Leterrier's 2010 version take a very different approach and reveal Medusa to be much more snake than human. Instead of having legs, she has a long writhing snake's body that tapers into a rattle at the end. Many camera shots in the films take close-ups of just the rattle or a small section of her body sliding between stone columns. This is scary in much the same way that seeing a shark fin disappearing beneath the waves near swimmers is scary, as it makes good use of animal dangers that are real. For all the availability of antivenins and medical facilities, the human brain cannot shake off millions of years of evolution. Snakes are still widely perceived with tremendous fear and filmmakers know it. This makes any close-up of slithering snake bodies writhing through rubble or tall grass frightening.

As for Medusa's ability to petrify in these films, there is something spine-tingling about it. The reasons are elusive. Nobody today should believe that anything could ever turn them to stone. Perhaps the terrifying element of the petrifying gaze is the fact that, subconsciously, there is a fear of physical petrification when faced with horrific circumstances, a deep-rooted and uneasy concern the body might not flee as it is supposed to when being attacked. Is this part of an ongoing awareness that shock induced by fear can literally paralyze and sometimes kill?

Although the fears of her gaze have not changed much, Medusa seems to have evolved, but not due to mutation or selective breeding. She has transformed with the fears of humanity. Once upon a time, she took shape from the frightening uncertainties presented by fossils and the threats posed by venomous snakes. Now, with fossilization well understood, some filmmakers are trying to utilize her femininity to make her a manipulating monster like the Sirens. Others are focusing on the reptilian aspects of the historic Medusa that are still scary and making as much of these traits as they can. Where her evolution will end is tough to tell, but her recent portrayals suggest she will become more and more serpentine.

Snakes have lost none of their fear factor. They played a pivotal role in Steven Spielberg's *Raiders of the Lost Ark,* crushed the life out of people in Luis Llosa's *Anaconda,* were the stars of David Ellis's *Snakes on a Plane,* and have even slithered their way into J. K. Rowling's *Harry Potter* books.

Aside from appearing as the basilisk that petrifies the students of

Hogwarts in Chris Columbus's *Harry Potter and the Chamber of Secrets*, the ever-feared snake is consistently present alongside the villain Voldemort in the form of Nagini, his companion. Nagini is dangerous, of seemingly human intelligence, and intriguingly . . . female. Moreover, in David Yates's *Harry Potter and the Deathly Hallows, Part 1*, Nagini is revealed to be able to take the form of a human woman so she can lure Harry into cramped quarters in an old house where he can be more effectively ambushed. Once Nagini has Harry where she wants him, she transforms, grotesquely, into her snake form and attacks. Harry is petrified with fear and barely survives. The snake's attack is one of the scariest moments in the film. Medusa in modern form?

4

The Mysterious Fathoms—Charybdis, Leviathan, Giant Squid, Jaws

"You're off the edge of the map, mate. Here there be monsters."

—Captain Barbossa, *Pirates of the Caribbean: The Curse of the Black Pearl*

At sea, leaving sight of the land is always unnerving. As the shore slips away and the black waves grow choppy, there is a peaceful solitude that goes hand in hand with a sense of foreboding. Of course, on modern vessels there are often radios, life rafts that automatically pop open if the boats are struck by rogue waves, and emergency beacons that will alert rescue teams if the ship goes down, but even so, these essential bits of safety equipment do little to assuage a primal fear of vulnerability associated with the sea. For ancient mariners, the ocean was a powerful and dangerous force.

It is fair to ask whether fear of the ocean is as irrational as fear of the dark. The fear of darkness stems from apprehension about nocturnal predators, but the fear often felt when walking through a darkened bedroom is unwarranted today, since not many people have a threat of nocturnal predators lurking by their nightstands. In contrast, there is something real and substantial about fears of open waters.

The main cause of death in people left adrift is hypothermia-induced drowning. In most parts of the world, the ocean is far colder than the toasty 98.6° F (37° C) that the body needs to survive, and there is no way to generate enough heat when submerged to keep treading water for very long. Even in seemingly warm waters, like those off the Greek coast, temperatures are cooler than the core body temperature and sap the body of its heat. After a while in the water, a swimmer's arms feel like planks of wood, the muscles lock, and he sinks into the dark depths where he drowns.

Life vests change things a bit. Instead of dying by drowning, death tends to occur by actual hypothermia. The body at first burns nutrients to generate heat for itself, but this is a lost cause. If the adrift individual were cooling down in a shallow bath, there might be some hope for the body to heat the surrounding water and fend off the cold, but not in the limitless ocean. Drifting people, even those in wet suits, inevitably lose all their heat. Their temperatures plunge, their respirations and heartbeats come to a halt. With no blood flowing, organs fail and they die.* It is a grim way to go.

Few monsters better represent fear of the water and all of its life-quenching properties than the fiendish Charybdis. Literally a living whirlpool that has a taste for human flesh, Charybdis is famously featured in Homer's *Odyssey.* As the hero Odysseus sails on from his visit to the sorceress Circe, she tells of dangers ahead.

She explains he must pass through a narrow strait with the bizarre

*Remember how *Titanic* ends? Yeah, that's what hypothermia-induced drowning looks like.

Scylla on one side and Charybdis on the other. "Charybdis sucks the dark water down. Three times a day she belches it forth, three times in hideous fashion she swallows it down again. Pray not to be caught there when she swallows down; Poseidon himself could not save you from destruction then."

Due to the dangers presented by the living whirlpool, Circe warns that it is better to lose six men than his whole crew and advises Odysseus to sail close to Scylla. In the end, this is exactly what he does: Scylla feeds, and Charybdis gets no further description.

Certainly a basic fear of water and drowning played a role in the creation of Charybdis, but the fact that a whirlpool is specified suggests that the Greeks actually knew what oceanic whirlpools were. That whirlpools are real was made painfully obvious to most people in the modern world shortly after the catastrophic tsunami struck Japan in 2011. As the waters that rushed inland receded, they collided with incoming waters. Like two surging rivers flowing next to one another in opposite directions, these rushing waters started to spin where they met and created a huge vortex, sucking ships into its center. Could the Greeks or their ancestors have seen something similar and been led to imagine there was a monster dwelling in the water?

As mentioned earlier in "It Came from the Earth," fossils of marine organisms scattered among terrestrial deposits along the coast of Greece make it clear that the Mediterranean was the site of historic tsunamis. Unfortunately, whirlpools created by tsunamis leave no fossil evidence, so it is impossible to know for certain if they actually did occur. However, since we know that tsunamis were striking the ancient Greek coast and since the *Odyssey* specifically describes a whirlpool monster, it does not seem like much of a leap to suggest that the Greeks, or their recent ancestors, saw at least one huge whirlpool that scared them senseless and made its way into their myths.

But even this examination of Charybdis's origin might be too simplistic. Homer specifically points out that the monster vomited forth water three times a day and sucked it back down again three times a day. This is odd, since tsunami-formed whirlpools are one-time events that ultimately vanish, as the Japanese tsunami-formed whirlpool did. What Homer's description suggests is a whirlpool formed by tidal activity.

Reasonably strong whirlpools of this sort—where water rushing out with the tide from one location encounters water rushing in with the tide from another—do exist in a few places along the coasts of Scotland, the United States, and Japan. Just as with tsunami-formed whirlpools, these interactions create a vortex. However, these whirlpools strengthen and weaken on a set schedule that runs like clockwork with the tides. Since there are typically two high tides and two low tides per day, it is a bit baffling that Homer describes Charybdis as sucking in water, presumably through a vortex, three times a day. However, in some parts of the world, including the Mediterranean Sea, there can be unusual numbers of tides per day, with six (three high and three low) a possibility. Alternatively, by saying "three times a day," Homer may have simply been referring to tides noticed during "daytime," in which case it would be common for at least one of the whirlpool formations at tidal sites to take place during the dark of night and thus not be seen.

Exactly which real-world tidal whirlpool Homer was considering when describing Charybdis is a difficult question to answer since the existence of whirlpools elsewhere in the world would have been unknown to the Greeks, and most regions of the Mediterranean do not have tides of any significant strength. A naval chart produced by the Italian government in 1881 includes a tiny whirlpool drawing just south of Sicily's Capo Peloro. It is labeled "Charybdis" and marked as a hazard. Another chart, created in 1823 by Captain W. H. Smyth of the Royal Navy, places a whirlpool drawing with the "Charybdis" label in exactly the same place. Yet another, made in 1810, places the monster slightly farther north. Was legend leading people to mark the monster on maps even if no real monster was there? Or were real navigational conditions once so terrible in the area as to warrant the making of a monster?

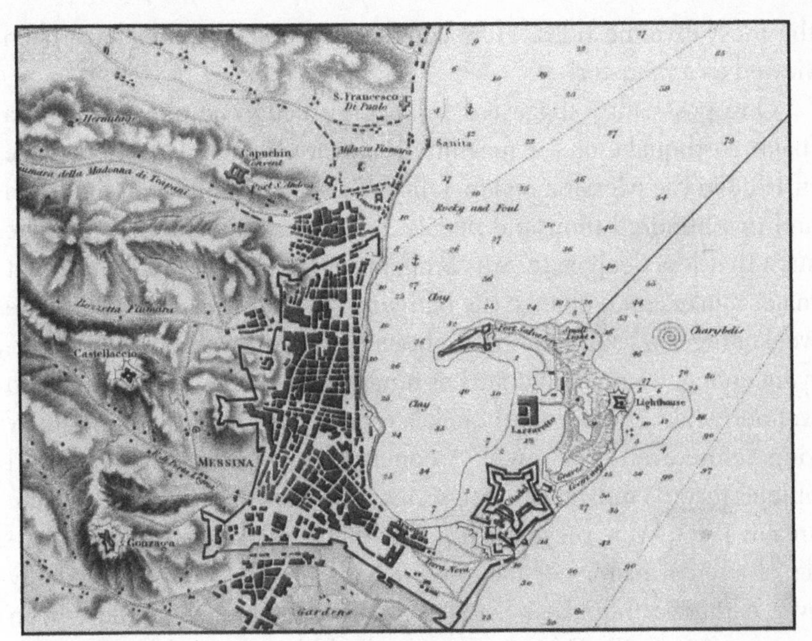

Plan of the Faro, or Strait of Messina, by Captain W. H. Smith, R.N. 1823. British Library.

Today, British admiralty charts mention whirlpools as regularly forming in the Strait of Messina, which separates mainland Italy and Sicily. Known locally as *garofali*, these whirlpools are actually tidal in nature because, while tides in the Mediterranean are very weak, the narrow strait amplifies the mild tidal effects that are present there. This amplified tide also runs across an unusual submarine ridge that allows it to sometimes drag up cold, dense water from deep below. After being dragged up and moved a short distance on the surface by the tide, this cold water quickly sinks back down to the depths and buoyant warm water swiftly rushes in to fill its place. This process is responsible for the creation of whirlpools. Moreover, the British admiralty specifically identifies one whirlpool, near Torre Faro on Sicily, as particularly large and permanent, and states that it is widely believed to be the Charybdis of Greek lore. What is odd is that this "Charybdis" is listed as hazardous only to small watercraft during the most extreme tides. How could something so minor have been viewed as a monster?

One possibility, suggested by modern oceanographers, is that a major earthquake of 7.2 magnitude on the Richter scale took place in 1908 in the Messina region, killed between one hundred thousand and two hundred thousand people, and altered the submarine ridge such that less deep water was brought to the surface. And if a recent major quake could change the bathymetry of the area and lead to the weakening of a whirlpool that eighteenth-century sailors thought worrying enough to note on charts as a monster, it is not unreasonable to wonder whether older earthquakes reduced the whirlpool's intensity long before even they saw it. A combination of geological clues and human records indicates that a series of powerful earthquakes hit the area in 1783. Were these earthquakes the first to weaken the Charybdis of ancient history, or had the monster already been weakened by even earlier earthquakes? For all we know, Homer may not have been exaggerating at all. Tides in the area could have once brought up so much cold water from the deep that they produced a whirlpool large enough to present a major threat to vessels of all sizes.*

Yet Charybdis is rare among sea monsters by being composed of water and killing people by drowning them. Far more often the danger presented by the ocean takes shape in the human imagination as something physical and predatory. Really, when left adrift in dark waters of seemingly limitless depth, there is nothing more horrible than feeling something swim just past the soles of your feet or, worse, having that something swim in for a bite. And it is from this fear that the legendary and vile Leviathan comes.

*Of course, a counter argument can be made that many of the "vessels" used during the classical period would charitably be called floating fruit baskets today, so the whirlpool might not have needed to be that much bigger to have been viewed as dangerous.

Biblical fears

Huge, hulking, and powerful, Leviathan is staggeringly different from Charybdis in having a physical form capable of swimming long distances and causing tremendous destruction. Unlike Charybdis, which occupies only two lines in the *Odyssey,* the biblical description of Leviathan in the book of Job is considerable:

> *Any hope of subduing him is false; the mere sight of him is overpowering . . .*
> *I will not fail to speak of his limbs, his strength and his graceful form.*
> *Who can strip off his outer coat? Who would approach him with a bridle?*
> *Who dares open the doors of his mouth, ringed about with his fearsome teeth?*
> *His back has rows of shields tightly sealed together;*
> *each is so close to the next that no air can pass between.*
> *They are joined fast to one another; they cling together and cannot be parted.*
> *His snorting throws out flashes of light; his eyes are like the rays of dawn.*
> *Firebrands stream from his mouth; sparks of fire shoot out.*
> *Smoke pours from his nostrils as from a boiling pot over a fire of reeds.*
> *His breath sets coals ablaze, and flames dart from his mouth.*
> *Strength resides in his neck; dismay goes before him.*
> *The folds of his flesh are tightly joined; they are firm and immovable.*
> *His chest is hard as rock, hard as a lower millstone.*
> *When he rises up, the mighty are terrified; they retreat before his thrashing.*
> *The sword that reaches him has no effect, nor does the spear or the dart or the javelin.*
> *Iron he treats like straw and bronze like rotten wood.*
> *Arrows do not make him flee; slingstones are like chaff to him.*
> *A club seems to him but a piece of straw; he laughs at the rattling of the lance.*

It is an unusually long description for a monster, but it is very much worth taking apart to understand how humanity moved from conjuring up a living whirlpool to a creature like this.

Consider "His back has rows of shields tightly sealed together; each is so close to the next that no air can pass between." While the image of an animal with shields for skin is vivid, it is hard for the mind not to wander to thoughts of reptiles when reading something like this. Whether Leviathan's skin is inspired by the large, hard scales of the crocodile or the scutes on a sea turtle's shell is tough to tell from the description. But anyone paddling along the Nile would have had a chance to tangle with crocodiles of substantial size, and anyone sailing near the Egyptian coast would have likely seen the backs of sea turtles making their way through the water. From just this description of Leviathan, it would seem that such reptile encounters found their way into the biblical texts.

At first glance, the description of sparks of fire shooting out of Leviathan's mouth seems the stuff of pure fantasy, but then follows, "Smoke pours from his nostrils as from a boiling pot over a fire of reeds." The critical thing to correct, of course, is that smoke does not actually come from a boiling pot. What is meant here is probably steam, though this makes little sense since the only features in the natural world that produce steam are geothermal in origin, and it would be difficult to mistake a volcanic eruption or a geyser for a sea monster. But steam and mist look awfully similar when seen from a distance, and whales, when they exhale through their blowholes, could easily be mistaken for releasing steam by early sailors who couldn't (or wouldn't) get close enough to feel that the steam was actually cool to the touch.

The possibility of a whale inspiring the Leviathan myth is supported by other parts of the biblical description. For example, "When he rises up, the mighty are terrified; they retreat before his thrashing" seems to be an account of a whale breaching from beneath the waves. And the arrows, clubs, and slingstones being unable to harm Leviathan further support this idea, since whales have incredibly thick skins and usually require harpoons to be killed.

Several whale populations might have inspired these elements of

Leviathan. It might seem sailors would have to leave the Mediterranean and head out into the Atlantic to see anything bigger than a dolphin, but this is not true. Even though whales are not often seen by locals today, there are a few whale populations that have adapted to survive in the Mediterranean. A population of fin whales, large and docile animals that passively feed off of plankton much as blue whales do, live in the region and are seen from time to time. More intriguingly, there is a sperm whale population, the huge carnivorous whale species featured in *Moby-Dick*, living off the shore of Zakynthos in the Ionian Sea. Modern biologists' realization that the whales of Zakynthos are sperm whales came as quite a shock, since male sperm whales are well known for making incredibly long migrations between the near-freezing waters found at the poles and the warm waters near the equator. As it turns out, the male sperm whales near Zakynthos don't migrate at all but instead stay in the warm Mediterranean water with the females and young of the population throughout the year.*

These whale populations were probably present between 1500 and 500 BC, when Leviathan took shape in biblical writings, since lots of pottery from that time depict monsters that mix whale characteristics with those of reptiles, hinting that the artists of these works had at least caught glimpses of whales. Moreover, archaeological studies have revealed a whale shoulder bone that was used as a chopping board in an ancient Greek marketplace around 900 BC, roughly a hundred years before Homer. While it is possible that the bone was collected from the coast of the Atlantic and brought to Greece, it is more likely that it was part of a whale's body that washed up locally.

*Hanging out in the Mediterranean year-round sounds like the easy life, but in fact these whales are among the most threatened in the world. There are no Icelandic or Japanese whalers to hunt them, but they are effectively living in a giant fishbowl (or perhaps a better metaphor would be a giant toilet). The Mediterranean has only a very small opening connecting it to the Atlantic, and not much water circulates between the two. This means that any pollutants produced by the huge (and often not especially conservation-minded) populations living around the Mediterranean stay in the sea for decades before circulating out. Exactly what long-term effect these pollutants are having on the whales is not entirely known, but it can't be good.

So Leviathan looks to have arisen from a mix of fears. Some sailors must have encountered large marine or river-dwelling reptiles and had no idea what they were. Others must have seen whales and been utterly terrified by their size, breaching behaviors, and misty exhalations. The result was the creation of a monster with mixed traits that blended features of animals that nobody could make sense of.

And Leviathan was only the first of many. In Greek mythology, the princess Andromeda was to be sacrificed to a flesh-and-blood sea monster named Cetus and saved only at the last moment by Perseus, who held up the severed head of Medusa to turn the monster to stone (in the film versions of this tale, *Clash of the Titans,* Cetus is renamed Kraken, but the monster is effectively the same). Artistic renditions of this vary, and many focus only on the curvaceous Andromeda with Cetus either in the distance or out of sight completely. One piece of art that does give the monster a lot of attention is a ceramic jar made around 510 BC. On it, the figure of a hero, probably Perseus, prepares to do battle with an enormous sea monster, probably Cetus (some suggest this is Hercules preparing to battle a sea monster, but nobody is sure). What is remarkable about this art is that there are so many sea animals drawn with such accuracy. The dolphins above and below Cetus, the octopus in the lower left corner, and the seal in the far left are all immediately recognizable. Cetus, on the other hand, is a mess. The monster has the highly maneuverable pectoral fins (the ones in the front left and front right of the body) that a whale has. However, the snakelike undulating body and the spinal frill clash with the idea of this beast being a whale. It also has large gill slits behind the jaw and an anal fin sticking out from the base of its body just a bit in front of the tail. These are not whale traits at all; they are fish characteristics that are easily seen on sharks, hinting that the artist had seen a shark at some point in time. Given the artist's superb drawings of the dolphins, seal, and octopus, it seems doubtful the drawing of Cetus was just the result of an animal like a sperm whale being drawn incorrectly. This is more likely to be an artistic attempt at creating something truly scary by combining the characteristics of multiple animals.

Hercules or Perseus and Sea Monster. Greek Caeretan black-figure clay vase, c. 530–520 BC. Stavros S. Niarchos Collection, Athens. Photo by Silvia Hertig.

And such mixed sea monsters did not stop being drawn with the end of the Classical era. In 1554, the Renaissance artist Titian painted this scene and again presented Cetus as a monstrous mélange.

In Titian's *Perseus and Andromeda,* Andromeda is fair, chained to a rock, and (as always) naked. Perseus, wearing winged sandals, is diving almost vertically through the air ready to do battle with the approaching monster. Cetus's cavernous mouth is open wide, and his body is covered in thick scales. The mouth could be inspired by whales, the scales by crocodiles or turtles, and the long, coiled serpentine body by eels or snakes.

While it is not clear exactly which animals inspired the beast in Titian's work, the scene is meant to convey great tension. With Perseus diving through the air and Cetus ready to strike, we fear for Andromeda. And this is important, because it points out that even in AD 1550, more than a thousand years after Leviathan and Charybdis took form, people still feared monsters of the sea.

The fear seems to have been ongoing because medieval maps

of the ocean are littered with drawings of strange and frightening creatures living far beyond the coast. Many of these resemble—at least partially—whales or snakes. But there are other beasts alongside these monsters that are truly foreign-looking things with long necks, toothy mouths, and large paddlelike flippers. There are no known animals that look like this in the oceans today, but this doesn't mean medieval sailors were just making things up. They may have been looking at fossils.

Perseus and Andromeda, by Titian. Oil on canvas, 1554–1556. By kind permission of the Trustees of the Wallace Collection, London. Art Resource, NY.

During the days of the dinosaurs, there were marine reptiles of huge size. Some were dolphin-like in morphology, but others had long necks and paddle-like fins. All had sharp teeth and must have puzzled those who found them. As the bones of these obviously aquatic animals turned up, assigning them to the category of "sea monster" was probably the most logical option that people had.

Although plesiosaurs have maintained a meager existence as monsters in Scotland, where the Loch Ness monster charade has persisted for decades, elsewhere, fear of them has waned alongside fears of whalelike and aquatic serpentine creatures. Little by little, maps of the oceans show ever fewer monsters until finally, by the twentieth century, sea monsters cease to feature on them at all.*

A big part of this sea monster extinction is probably linked to believability and human domination of the seas. With major steps forward in marine biology and paleontology during the Victorian era, whales started being identified as the docile animals that they are and plesiosaurs started being identified as extinct (except among the Scots). But these developments in science did not bring an end to sea monsters; they just forced them to evolve into more acceptable forms and move to more believable environments.

Still beneath the waves

In Jules Verne's *20,000 Leagues Under the Sea,* published (in French) in 1870, the sea monster that attacks Captain Nemo's submarine is a huge squid or octopus (depending upon how the French is translated). Compared to plesiosaurs and Leviathan-like creatures, here was a monster that many believed really could exist. Giant tentacles of seemingly enormous squid have occasionally washed up on beaches alongside the Pacific and Atlantic Oceans for centuries. Further evidence can be seen on sperm whales that are now known to eat enormous squid and sometimes get huge sucker marks scarred across their bodies.† Yet because squid do not breathe air, as whales

---

**National Geographic* sent an expedition to the loch in 1977 that overturned every rock, scanned every ripple, and monitored the entire ecosystem so closely that if a monster did exist, it would have been found. Nothing turned up. Years later it was revealed that the key monster photograph released decades earlier was a fraud.

†Some marine biologists have taken on the task of studying the contents of dead sperm whale stomachs. It is a dreadful (and smelly) activity, but it has proved, without any doubt, that sperm whales do eat giant squid.

do, and seem to live their entire lives in the deepest depths of the ocean, exactly what they looked like and how they behaved were left to the imagination.

We now know more about giant squid because several full-bodied individuals have been found and studied. From the ends of their wickedly barbed tentacles to the tops of their heads, giant squid really are giant, measuring 43 feet (13 meters) in length. Moreover, there appears to be a second, even larger, species in need of study, that of the colossal squid that some estimates suggest can reach 49 feet (15 meters) in length. Yet how these species behave remains a mystery, since they live in such isolated environments and are rarely seen alive.

Of course, not having easy viewing access of giant and colossal squid has hardly stopped scientists from trying to better understand them using other methods. In 2010, a team of a researchers published a paper in the *Journal of the Marine Biological Association of the United Kingdom* in which they calculated the metabolic activity of other, better-known, squid species and scaled this information up to apply to an animal of the colossal squid's size dwelling in the extreme cold of the deep ocean.

Based upon their numbers, the team, led by Rui Rosa at the University of Lisbon, worked out that the colossal squid used far less energy than similarly sized animals of the deep ocean, like sperm whales. With such low energy demands, the researchers argued that the squid were probably not functioning as pursuit predators that chased their prey around but instead as ambush predators that would just snatch the occasional (large) fish out of the water.

The media took this finding to indicate the colossal squid did not deserve its reputation as a monster, but that opinion may have been made without properly digesting the research. Does an animal need to be a pursuit predator to be worth fearing? Sure, some great cats are frightening hunters that will chase people down and eat them given the right conditions. But crocodiles and pythons are ambush predators with low energy demands that rush out from hidden locations to grab prey. If anything, the ability to hide and strike without warning is what makes these animals so inherently terrifying. So if this is true of reptiles, why not also of squid?

To date, there have been no trustworthy accounts of giant or colossal squid attacking humans or ships, but there are accounts of smaller squid attacks. While making a documentary film for the Public Broadcasting Service in 1997, five-foot (two-meter)-long Humboldt squid in the Sea of Cortez "mugged" a diver. The account, described on the PBS website by the cinematographer, is startling: "Three squid had taken his collecting bags and bottles, his dive computer and the gold chain from around his neck. These squid had quarter-sized suckers, lined with teeth for tearing apart their prey, and Alex was left with a series of round, red scars circling his neck. Adding insult to injury, the squid dragged him down very deep before letting him go." Certainly, if the Humboldt squid in the Pacific Ocean is any indicator, it appears reasonable to argue that giant and colossal squid present at least some danger to divers. Even so, the remoteness of the environments that these squid live in and the fact that so few documented attacks are associated with these animals keeps their monster status meager at best. To be threatened by them, people have to really work hard at getting close. Sharks, however, are another matter entirely.

Peter Benchley's novel *Jaws* and Steven Spielberg's movie based on the novel presented the great white shark as the stuff of nightmares. A big part of this monster's creation was inspired by a series of real shark attacks that took place off the coast of New Jersey during the summer of 1916. Between July 1 and 12, five people were attacked by a shark and four died. On July 14, shortly after the surge of deaths, a 7-foot (2.4-meter) great white shark was killed nearby and the attacks stopped. Whether the attacks were those of a single rogue shark or a sudden burst of attacks from multiple sharks driven to feed upon humans because of changing environmental conditions was never determined, but the result was the birth of a new sea monster. Or at least that is how it would appear.

Jaws spawned a series of financially successful sequels and cultivated a major industry based upon scaring the hell out of people with shark footage, but for the popular reaction to *Jaws* to be so powerful, there had to be something more to the fear than just the 1916 shark attacks. Shark attacks taking place around the world have been recorded by the Florida Museum of Natural History for more than a hundred years and the records hint at a disturbing trend. From 1900 to 1910, only 25 shark attacks are recorded as having taken place around the world. During the 1970s, the decade when *Jaws* was published, the number was around 140. From 1990 to 2000, the number was close to 500.

It is doubtful that these figures accurately represent an increasing hunger for human flesh among sharks. Reporting of shark attacks has improved, and an attack in a remote location is much more likely to find its way to the Florida Museum's records today than it would have been in 1905. In addition, the human population has dramatically increased in size, and with more people around, there are more swimming bodies to be attacked. Finally, the popularity of beach holidays has exploded alongside activities that draw people to the water. Surfing, scuba diving, kayaking, and snorkeling were not widely popular (or in some cases even invented) when the Florida Museum started collecting data. Their spread around the world has simply put more humans in the water and led to more chances of encounters with sharks.

Whatever the reason(s) for the increased reporting of shark attacks, there is a snowball effect at work too. With increased reports there is increased media coverage, and with increased media coverage there is increased fear. Between the 1950s and 1960s, the Florida Museum reports reveal that shark attacks made one of their biggest leaps of the century, jumping from roughly 150 reported during a decade to 250. Even though Peter Benchley explained during interviews that *Jaws* was inspired by the 1916 attacks, one has to wonder if a pulse of shark fear during the decade when he wrote the book also played a role.

Yet it would be wrong to argue that the past century of increasing shark attacks is solely responsible for the creation of the shark as a modern monster. While psychological literature on human fear of water predators is nowhere near as extensive as it is on human fear of snakes, humans evolved in Africa and there are many sharks along some sections of the African coast. Were interactions taking place?

A 2007 archaeological study led by Curtis Marean at Arizona State University and published in *Nature* revealed the discovery of 160,000-year-old shells and stone tools at the South African coastal site of Pinnacle Point. It is located roughly halfway between where Cape Town and Port Elizabeth are today and is believed to be one of the earliest locations where humans were regularly turning to the ocean for food. Why they were suddenly seeking out shellfish is a matter of debate. One argument is that as the climate became cooler and drier at that time in history, food on land became more scarce, and other food sources needed to be found. Whatever the reason, it is clear from the archaeological evidence that the species *Perna perna*, better known as the brown mussel, was collected quite a lot by the humans who lived at this location.

We don't know the extent to which humans were actually entering the water, but they were spending a good deal of time next to an area that the Florida Museum database shows to be one of the shark attack capitals of the world. Compared to nearby Mozambique, Tanzania, and Madagascar, which have recorded 11, 4, and 3 attacks respectively since 1828, South Africa has logged 223. That is a big difference. And if modern humans are easily attacked in the area, why not also shellfish collectors seeking to get a particularly large animal sitting on a rock just a short distance offshore?

The reality is that the brown mussel, which was being collected so intensely 160,000 years ago, is found at the very top of the rocky intertidal zone, and sharks rarely swim into such shallow waters. But Marean points out that by 110,000 years ago, humans were foraging lower down in the rocky intertidal zone and also in sandy beach areas where sharks would have been more common and had the opportunity to attack. It is impossible to know for certain if such attacks were frequent, since fossil evidence of shark attacks on people (human bones with shark teeth stuck in them) are extremely rare in the fossil record.* Even so, the evidence hints that human ancestors who tried collecting food along the coast might have sometimes ended up as shark bait. It would seem that as civilizations graduated from just collecting shellfish to fishing with spears, humans and sharks came into even closer contact, since sharks are strongly attracted to the presence of blood in water and would most certainly attack a human who tried to keep them away from a freshly speared fish.

So, with all of this in mind, if studies suggest that the threat presented by snakes has driven evolutionary forces to select for humans who notice and fear snakes instinctively, have hardwired fears of dangers in the sea similarly evolved as the result of historic shark attacks on humans? It seems a reasonable assumption to make, but there is a problem. If the fear of being eaten by sharks is deep-seated and has been selected for by evolution, why does the shark as a monster not emerge until the 1970s in *Jaws*?

Sharks attack unseen. The vast majority of modern attacks are on limbs sticking off surfboards or legs hanging down beneath the water. Some sharks bite and let go, allowing their prey to bleed to death so they can then eat without worrying about being harmed by a struggling animal. This is the case with great white shark attacks

---

*Here's the logic. Shark attacks are rare to begin with. If they did take place in ancient times, the body would not have been recovered by terrified viewers of the attack but rather be left as fish food. In either case, human bones are extremely unlikely to fossilize under such conditions. Incidentally, in 2009, a paleontology team discovered shark teeth wedged into the bones of a plesiosaur (one of the huge marine reptiles discussed earlier that are often associated with Loch Ness). The teeth were of different sizes, indicating that the enormous reptile was attacked by numerous sharks. They argued that it was quite possibly the earliest evidence ever found of sharks entering a feeding frenzy. And because the attacks were made on an animal with big bones that did not break easily, the evidence actually got recorded. The bones of a human caught in a shark feeding frenzy would likely be obliterated.

against elephant seals on the California coast. However, some sharks attack by grabbing prey on the surface, crushing it with their jaws, and dragging it down deep so that it will drown. Witnesses of both types of attacks would see almost nothing other than some thrashing on the surface, blood in the water, and a screaming victim who would sometimes sink down into the depths and never be seen again. The details of the creature making the attack would largely be left up to the imagination. Moreover, sharks do not tend to beach themselves nearly as often as whales do, meaning ancient people would have had fewer opportunities to examine these predators from the safety of dry land. Really, the only reason people today have such a good understanding of what sharks are and how they attack is thanks to modern devices like scuba masks, goggles, and underwater cameras that allow shark activities to be widely seen.

So if sharks were making historic attacks on humans, and if these attacks were almost entirely invisible, did this lead people to instead point at the weird whirlpools, whales, reptiles, and fossils that they could easily see and use the traits of these things to shape their vision of what sea monsters were? It seems a reasonable possibility. Indeed, if this is so, the terror responsible for the creation of the monster in *Jaws* is really no different from the terror behind Leviathan. The fear is the same; the form that the fear takes is the only thing that has changed.

Where fear of sea monsters is headed is a challenge to predict. Marine biologists and conservation groups rightly point out there is little evidence that sharks can develop a taste for human flesh and become "man-eaters." These same groups agree that shark attacks are mostly accidents where the shark does not actually intend to attack a human but instead mistakes a human for one of its typical prey species. This, in theory, should calm concerns and reduce shark fear. Even so, media events portraying sharks as dangerous, like the Discovery Channel's Shark Week, still gather a lot of interest. So too do thrill activities like shark diving, where food is intentionally thrown into the water to attract sharks while divers are below the surface in a cage.* But there is a notable reversal worth mentioning.

In 1989, the director James Cameron made *The Abyss*, a film about a military submarine that sank under puzzling circumstances and the presence of a mysterious form of life in the deep ocean. Although the creatures dwelling in the abyss are initially perceived as threatening, it is the people who ultimately prove to be the true danger. The film ends with the deepwater creatures saving the protagonists and the human villain being crushed by water pressure at depth. While not really a monster movie, the story is notable for its portrayal of the dark depths as places of wonder.

That this clashes with instinctive fears of dark or murky water and the predators hiding within is unmistakable. That it runs alongside some astounding recent research showing deep ocean environments to be special places with remarkable life-forms in need of protection makes such a story all the more worth telling. The critical question is if people can overcome the innate fears that may have helped keep their ancestors alive.

Whether fear or reason wins will ultimately depend upon what stories get told. For the sake of the animals in the deep oceans, hope for more tales like *The Abyss*.

---

*There is a chance that this tourist activity is actually teaching sharks to associate human divers with food and potentially increasing the number of shark attacks in regions where this activity is happening. Whether sharks are smart enough to make this association remains to be determined, but it is an idea that should at least give would-be shark divers pause.

5

Of Flame and Claw—Dragons

"They're seriously misunderstood creatures."
—Rubeus Hagrid, *Harry Potter and the Goblet of Fire*

Dragons are among the world's most enduring monsters. They appear early in Babylonian myths about the great wars fought between their gods, then crop up again in tales of the Greek hero Jason and the witch Medea. Later, they appear in medieval lore and are famous for doing battle with the likes of Saint George and Saint Margaret and for breathing blasts of deadly fire upon warriors like Sigfried and Beowulf. Dragons have even made it to the modern day, attacking Bilbo Baggins and chasing after Harry Potter on his broomstick. Yet in spite of their long life span as monsters, and the widely varied stories that they have appeared in throughout the ages, dragons are remarkably consistent in form.

Tiamat. Cylinder seal impression. Neo-Assyrian, 900–750 BC.
© The Trustees of the British Museum. All rights reserved.

No fuzzy or remotely mammal-like dragons have ever been described. Even the most ancient of these monsters are covered in reptilian scales. According to Babylonian mythology, there was an ancient conflict between the great gods Apsu and Tiamat and their children. This conflict led to Apsu being killed and Tiamat growing very angry. To avenge her partner's murder, Tiamat transformed herself into a serpentine creature with horns and a wriggling body. Exactly what this creature looked like is not clear from ancient writings, but artwork offers a hint. A cylinder seal at the British Museum shows Tiamat with a horned head, a lengthy tongue, tiny forelimbs, and a very long body.

The similarity to a serpent is obvious, and it seems fair to ask if snake fear was playing a role in inspiring the form Tiamat took. The Babylonians certainly would have had some exposure to venomous snakes and, when they were trying to come up with a frightening form for their god to take, simply settled on one belonging to a dangerous animal in their environment. Yet the horns are a bit of a mystery. Goats have horns, and there were certainly goats around Babylon, but goats do not traditionally qualify as scary animals. However, a look at venomous snakes provides a possible answer.

One of the most dangerous snakes in southern Europe and the Middle East is a species known as *Vipera ammodytes,* commonly known as the sand viper. It has long fangs that can readily puncture human skin, a foul temper, and potent venom. Whether this specific snake is related to the dragon that Tiamat transformed into is, at first, disputable. Yet if the scales on its head are taken into account, they hint at a connection, since scales just behind the sand viper's eyes grow larger than those on the rest of its head and look like horns.

In other parts of the world, the connection between snakes and dragons is also strong. In the story of the Golden Fleece, the hero Jason goes with the witch Medea to collect the fleece from a dangerous creature. What this creature is exactly is also not obvious. The Greek poet Pindar, who lived during the fifth century BC, wrote, "For the fleece was laid in a deep thicket, held within the fierce jaws of a ravenous dragon, far surpassing in length and breadth a ship of fifty oars." Yet according to the poet and scholar Apollonius Rhodius, "The fleece is spread on top of an oak, watched over by a serpent, a formidable beast who peers all round and never, night or day, allows sweet sleep to conquer his unblinking eyes."

The Greek texts call the monster *dracos*. This is the word from which the Anglo-Saxon word "drakan" and the modern words "drake" and "dragon" are thought to come, but the ancient Greek term is ambiguous. It was also the word for snake. This is why some English translations of the various historical accounts call the creature Jason tangled with a serpent.

To complicate matters, the art associated with Jason and his quest is also inconsistent. On one iconic Greek jar (made at an unknown date during the Classical period, 500–300 BC), Jason is reaching for the Golden Fleece as a snake rears up from behind the treasure preparing to strike. Yet on a Greek plate made sometime between 500 and 450 BC, the monster guarding the fleece dwarfs Jason as it vomits him up after grabbing him with its rows of sharp teeth.* It

*Nobody understands the version of the story told on this plate. The writings have all been lost. Presumably Medea (or possibly Athena) cast a spell that made Jason taste so terrible that the monster vomited. But that is just a guess.

has a noticeably snakelike body but otherwise looks a lot like what modern audiences would consider a dragon.

Jason about to Steal the Golden Fleece, attributed to the Orchard Painter. Greek, terra-cotta column-krater, c. 470–460 BC. © Metropolitan Museum of Art, New York. Art Research, NY.

Jason and the Dragon, attributed to Douris. Greek ceramic kylix, 500–450 BC. Museo Gregoriano Etrusco Vaticano, Vatican City. Art Resource, NY.

Stories of other Greek dragons further complicate the situation. In Euripides' play *Medea,* which describes Medea's activities after she helped Jason steal the Golden Fleece, she runs into trouble with the law in Corinth after she murders its king and his daughter. To make her escape, she uses her magic to conjure dragons. Yet these creatures can fly. A text that has historically been attributed to Apollodorus describes them as "winged dragons" pulling a chariot. Yet the Roman poet Ovid writes that after pulling Medea's chariot over great distance, the dragons "sloughed their aged skins of many years." They had wings, a distinctly nonserpentine characteristic, but shed their skins just as snakes do. This hints at snakes being key for dragon inspiration, but venomous snakes help to explain only some of the physical appearance of dragons. When it comes to their size, venomous snakes don't provide much of a model at all.

Dinosaurs and their kin make an obvious choice for the ancestors of dragons since they were often very big and left behind skeletons that appear distinctly dragonlike in form. However, dinosaurs present a problem because their fossils are not found in Greece and are rare throughout much of the Mediterranean. Even so, some Greeks sailed to far-off places and may have encountered large reptile skeletons on their travels.

Pytheas, a Greek explorer, made it to Britain around 325 BC and seems to have traveled as far north as Scotland and as far west as Cornwall. If he really did make it that far west, he would have sailed along the Dorset coast, an area littered with the fossils of giant marine reptiles.

Toothy and fierce-looking plesiosaurs, pliosaurs, and ichthyosaurs are all commonly found eroding out of the rocks in this region. In fact, so many of these animals have been found in the area that it has come to be known as the Jurassic Coast. Pytheas would very likely have stopped along these shores and possibly encountered fossils along the way. And that is just the story of Pytheas. Himilco, a Carthagian explorer who was active roughly a hundred years earlier than Pytheas, journeyed to Spain, France, and England, and may have had similar encounters.

If a plesiosaur neck, or the neck of some closely related species, was found embedded in rock, how would people living long ago have made sense of it? Would they have come to the conclusion that it was the neck of a giant snakelike monster? Most large marine reptiles had sharp teeth too. These teeth were not anything like the fragile, venom-injecting fangs found in snakes but more robust and dragonlike. If just the head and neck of one of these animals were found, it seems plausible that this could have raised fears that such huge carnivores really existed. But there is more.

In *The Dragons of Eden,* Carl Sagan argues that we should consider the ancient environment where early mammals first lived when thinking about the fears that humans have. Mammals evolved during the Mesozoic era, when reptiles ruled the world. Dinosaurs were everywhere, and the few mammals that were around were tiny rodentlike creatures that spent most of their time scurrying away from death in the form of *Velociraptor* and juvenile tyrannosaur teeth. For this reason, Sagan suggests that the appearance of any mammalian genes that coded for an inherent fear of and respect for large reptiles would have led to increased survival, and this fear would have been driven by natural selection to become common throughout the early mammal population.*

In recent years, Sagan's theory has gained a lot of support. A paper published in *Proceedings of the National Academy of Sciences* in 2011, which documented the interactions between preliterate Philippine hunter-gatherers and giant pythons, found that fifteen of fifty-eight people (26 percent) in the local population had been attacked by huge pythons during living memory. Most exhibited substantial scars from the bites made during the snakes' attempts to grab hold of them. Moreover, a startling 16 percent of the people knew at least one tribe member who had been killed by the reptiles. The researchers behind the study counted up a total of six fatalities in the past generation.

One man named Dinsiweg was found inside the body of a large snake that was killed and cut open by his son in 1940. Another, a twenty-five-year-old man named Diladeg, did not return home from hunting one day and was found the following afternoon crushed to death in the coils of a python. Two children had been consumed by a python that slithered into their home just before sundown in 1973, and in the same year a woman named Pasing died from an infection that developed from a python bite.

The researchers learned from the people's stories that most

---

*Sagan also points out that one of the sounds that humans around the world most commonly use to command attention is *sh* and asks if we can realistically view it as just chance that this happens to be a sound made by many snakes.

attacking pythons were between 16 and 32 feet (5–10 meters) long, tended to make ambush attacks when men were walking through dense rain forest seeking game, and were most often fended off with large knives. What struck the scientists as astonishing, however, was that even with the availability of metal knives, fatalities still often occurred. They reasoned that in the days before the natives had metal knives, fatalities would have been far higher than the 9.6 percent of the population per generation that the giant snakes currently claim. They don't give a proposed percentage, but even if it were 14 percent, that would be quite a lot. As a comparison, if 14 percent of the U.S. population were picked off by pythons in a generation, that would be more than 40 million people.

The shocking thing about this recent python study is that it provides good evidence that huge, nonvenomous snakes have been feeding on humans for a long time. So it raises the question: Were the depictions of the Babylonian Tiamat the result of sand viper characteristics being mixed with those of giant constrictors? Probably.

According to reports made by several Roman historians, in the midst of the First Punic War in the late summer of 256 BC, Roman troops heading toward Carthage (in Tunisia) were attacked by a great serpent when they came to a river. The following is an account by the historian Orosius:

> Regulus, chosen by lot for the Carthaginian War, marched with his army to a point not far from the Bagradas River and there pitched his camp. In that place a reptile of astonishing size devoured many of the soldiers as they went down to the river to get water. Regulus set out with his army to attack the reptile. Neither the javelins they hurled nor the darts they rained upon its back had any effect. These glided off its horrible network of scales . . . so that the creature suffered no injury. Finally, when Regulus saw that it was sidelining a great number of his soldiers with its bites, was trampling them down by its charge, and driving them mad by its poisonous breath, he ordered ballistae [wooden devices similar to

> catapults] brought up. A stone taken from a wall was hurled by a ballista; this struck the spine of the serpent and weakened the constitution of its entire body. The formation of the reptile was such that, though it seemed to lack feet, yet it had ribs and scales graded evenly, extending from the top of its throat to the lowest part of its belly and so arranged that the creature rested upon its scales as if on claws and upon its ribs as if on legs. But it did not move like the worm, which has a flexible spine and moves by first stretching its contracted parts in the direction of its tiny body and then drawing together the stretched parts. This reptile made its way by a sinuous movement, extending its sides first right and then left, so that it might keep the line of ribs rigid along the exterior arch of the spine; nature fastened the claws of its scales to its ribs, which extend straight to their highest point; making these moves alternately and quickly, it not only glided over levels, but also mounted inclines, taking as many footsteps, so to speak, as it had ribs. This is why the stone rendered the creature powerless. If struck by a blow in any part of the body from its belly to its head, it is crippled and unable to move, because wherever the blow falls, it weakens the spine, which stimulates the feet of the ribs and the motion of the body. Hence this serpent, which had for a long time withstood so many javelins unharmed, moved about disabled from the blow of a single stone and, quickly overcome by spears, was easily destroyed. Its skin was brought to Rome—it is said to have been one hundred and twenty feet in length—and for some time was an object of wonder to all.

The "serpent" is clearly a snake of some sort. It has ribs but no feet, moves sinuously, and glides over the ground. The creature's "poisonous breath" can be explained by a foul smell from the river and, even though no snakes can eat multiple adult humans in such a short time, it is possible that numerous men were bitten and drowned by a single large snake. However, no snakes attain 120 feet (36 meters) in length. *Titanoboa,* a python relative from the days of the dinosaurs, grew to 48 feet (15 meters), but snakes of such huge size have never coexisted with humans. Were the authors of the story just making this up? Or were they somehow getting their measurements wrong? In 2004, Richard Stothers at the Goddard Institute for Space Science* wrote about the Bagradas River incident in the journal *Isis,* proposing that the ancient reports of this snake mixed up 120 feet in length with the figure of 120 rib pairs (which the snake could have had). Moreover, he argued that even though large constrictors are not found north of the Sahara today, Pliny the Elder reported large snakes sometimes swimming in groups across the Red Sea from Ethiopia to Arabia. If this were true, he reasoned, they might have made it to the Bagradas River as well.

All of this hints that classical civilizations may have been encountering some very large constrictors in northern Africa, and when these stories are coupled with the hundreds of people who have died in living memory in the jaws of crocodiles, Sagan's theories make a lot of sense. There might really be an ancient, genetically based fear of all reptiles present in mammals that is still being selected for by evolutionary forces. It is an intriguing idea to be sure, and one that could easily work in concert with fears of snakes and ancient reptile bones. Yet neither giant constrictors nor marine reptiles explain the fact that dragons ultimately evolved the ability to breathe fire.

A heated question

In the medieval period, dragons begin to blast flames from their mouths. In Geoffrey of Monmouth's *History of the Kings of Britain,* completed in AD 1136, an early king named Vortigern is desper-

*Aside from being a world expert on the structures of stars, Dr. Stothers was a remarkable historian and evolutionary biologist.

ately trying to build a fortified tower in what seem to be the hills of Wales. Yet every time his men lay down stones to build the tower, the ground trembles and the structure is destroyed. Searching for answers, Vortigern turns to his wise men, who advise him to find a boy without a father and pour the child's blood over the ground to calm it so the stones can be properly placed. Vortigern sends scouts out to find such a child, and they ultimately discover a boy dwelling in what is today the Welsh town of Carmarthen being ridiculed by other children as a bastard. Bingo.

Upon learning how his blood is to be used, the boy tells Vortigern that his advisers have got things wrong. He argues that there are dragons beneath the building site guarding the land, that the ground cannot be built upon, and that spilling his blood will do nothing to solve the fortification problem. When the king orders the earth dug up, dragons are indeed discovered in "hollow stones" fighting with one another and "breathing out fire as they panted." Vortigern promptly dismisses his old aides and declares that the boy shall become his new adviser. The boy goes on to support Vortigern as well as his successors Aurelius, Uther, and, eventually, Arthur. He is, of course, the legendary Merlin, and revealing the dragons underground is his first demonstration of magical ability.

Even before the tales of Merlin and King Arthur, dragons breathe fire in the epic poem *Beowulf*. The story, first told at some point between AD 800 and 1100, features a dragon that starts setting fire to everything in the region after a valuable cup is stolen from a treasure trove it is guarding inside a burial mound. Beowulf, known in the poem as lord of the Geats, goes to fight the dragon and faces dangerous flames.

> The hoard-guard heard a human voice; his rage was enkindled. No respite now for pact of peace! The poison-breath of that foul worm first came forth from the cave, hot reek-of-fight: the rocks resounded. Stout by the stone-way his shield he raised, lord of the Geats, against the loathed-one; while with courage keen that coiled foe came seeking strife. The sturdy king had

> drawn his sword, not dull of edge, heirloom old; and each of the two felt fear of his foe, though fierce their mood. Stoutly stood with his shield high-raised the warrior king, as the worm now coiled together amain: the mailed-one waited. Now, spire by spire, fast sped and glided that blazing serpent. The shield protected, soul and body a shorter while for the hero-king than his heart desired, could his will have wielded the welcome respite but once in his life! But Wyrm denied it, and victory's honors.

Beowulf is mortally wounded while fighting, but his ally, the earl Wiglaf, comes forth and saves the day: "'Twas now, men say, in his sovran's need that the earl made known his noble strain, craft and keenness and courage enduring. Heedless of harm, though his hand was burned, hardy-hearted, he helped his kinsman. A little lower the loathsome beast he smote with sword; his steel drove in bright and burnished; that blaze began to lose and lessen." The presence of so much fire breathing in medieval dragon lore suggests that the trait was important and connected to something that people were genuinely scared of. But what?

Humans have been using fire for more than four hundred thousand years* and, as such, there has been a long time for people to develop a healthy sense of respect for the damage that fire can do. Long before fire came under our control, it was present in the environment in the form of forest fires spawned from lightning strikes. When animals are exposed to fire and smoke, they universally flee, so it makes sense for a certain level of inherent fire fear to be deeply seated in the human mind.

Forest fires, however, do not create the illusion of anything

*Exactly when fire started regularly being wielded by humans is a subject that can lead to fierce shouting matches at paleontological and archaeological conferences, particularly around the bar. Unlike stone tools, which hold up pretty well over thousands of years, the ash and charcoal that are often the only remains of ancient fires are easily destroyed by the elements and rare to find in the fossil record. Thus the shouting.

breathing fire, so it is unlikely that these fires or even out-of-control campfires inspired the dragon's breath. With this in mind, it is worth thinking about situations where fires can seem to be alive.

Fire needs only two things, a fuel to burn and oxygen. Deny it either of these things and it eventually dies. In the case of wildfires, if a fire were to reach an area where combustible material was found in large amounts, like a meadow filled with dried grass, the fire would roar to life. If it were to reach a vent of natural gas, it could easily cause an explosion that would have scared ancient people senseless.

As mentioned with Chimera, natural gas vents do exist in some places on Earth and may be connected to fire breathing by being associated with some of the same scents as those emanating from tar deposits. So it is possible the concept of the fire-breathing dragon stems from the same environmental conditions responsible for Chimera. However, dragon legends from the Middle Ages onward frequently place dragons in subterranean locations and usually describe them as guarding treasure. Beowulf's dragon is living in a burial mound where ancient valuables are hidden away, and Merlin's dragons are living in stones beneath the foundation of Vortigern's castle. Smaug, from *Lord of the Rings,* lives under the Lonely Mountain and has so many coins in his trove that many have become embedded in his underbelly. Even J. K. Rowling presents a dragon in *Harry Potter and the Deathly Hallows* guarding treasure deep below the wizard bank Gringotts. All of this hints that surface features, like the natural gas vents in Turkey, are probably not closely associated with these monsters.

Is flammable gas found underground? Oh yes. In fact, it is much more likely for miners to run into explosive gas than it is for people walking around aboveground. Moreover, miners need light to see what they are doing and have a long history of bringing open flames with them into the depths.

Imagine some ancient miners digging down in search of valuable resources. While holding torches to see what they are doing, the miners chip away at some rock and suddenly open up a vent that releases a burst of gas at them.* A terrible fire would light up the cave, seemingly "breathing" forth from behind the rock. It would not be much of a stretch for people seeing such a sight, who had no understanding of combustion, to believe that there was a monster hiding behind the rocks attacking them with a blast of flame.

This scenario explains the presence of dragons underground, explains their ability to breathe fire, and can even explain why people perceived them as large and powerful, since a really big combustion event could easily sound like the roaring of a huge beast. Indeed, the dragons that Vortigern's men find with Merlin's guidance below the hills of Wales could have been sources of flammable gases released by digging and ignited by torches. Moreover, Wales had vast coal deposits in its hills in the days before the Industrial Revolution. Based upon what we know today of coal deposits, we can guess that the gas in question was either methane, which can both suffocate miners and blow them up, or hydrogen sulfide gas, which stinks like rotten eggs, is poisonous, and is also explosive.

One has to wonder if Merlin caught the scent of methane or hydrogen sulfide gas in the air, knew about the chemistry of coal mines, told Vortigern to have his men dig because he figured the chances were good that their torches would light something up along the way, and assumed they would view his ability to predict such explosive fire as magic. Moreover, not only does Merlin's hometown of Carmarthen sit right next to the Welsh coal fields but the type of coal in the area and the geologic setting are both highly conducive to coal gas explosions. If true, it seems appropriate to consider Merlin more of a geologist than a wizard.†

But as clearly as gases associated with coal can explain the dragons in the Merlin story, *Beowulf* presents a different situation. In

*Boom!

†Arthur C. Clarke said, "Any sufficiently advanced technology is indistinguishable from magic." It would be ironic if one of history's most famous wizards was, in fact, a scientist in disguise.

that tale, the dragon awakens and starts setting fire to the land after a valuable goblet has been stolen from a treasure the dragon guards in a burial mound. In this case, it is not mining that awakens the dragon but thievery. How a theft could possibly start a fire might seem vexing, but fortunately two researchers, Elizabeth Barber, a linguist and archaeologist at Occidental College in Los Angeles, and her husband, Paul Barber, a research associate at the Fowler Museum of Cultural History at the University of California, Los Angeles, put their minds to this task and proposed a rather brilliant solution at the Fifteenth Annual Indo-European Conference held at UCLA in 2003.

In analyzing the *Beowulf* text, they noticed there are only six concrete elements present in the dragon battle of the story. First, someone takes a cup from an old burial site. Then fire erupts and spreads. Near the entrance to the burial site, Beowulf stabs at the source of the flames but is unable to kill the creature blasting the fire out. The flames are foul-smelling. Beowulf's ally Wiglaf later stabs the monster and this time the flames weaken and die out. Finally, when Wiglaf goes inside the burial site, the line reads, "No vestige now was seen of the serpent: the sword had ta'en him," suggesting that Wiglaf's blade annihilated the monster, leaving behind no dead dragon to describe.

The Barbers argue that there is a lot of truth to the tale. Neolithic and Iron Age people in Europe often built chambers for their dead, where they placed many valuables, including dead pets and horses, for their friends and relatives to carry with them into the afterlife. These burial chambers frequently had mounds heaped over them, and the sediment was often so fine that it effectively sealed the tomb aerobically from the surface environment. Archaeologists have noted in recent years that the sealing effect of these mounds was so significant that methane-producing bacteria, which normally live in the oxygen-poor human and animal intestines, are able to expand their range and consume much more than waste inside the gut. As they spread, they produce lots of methane, which builds up inside the small, well-sealed, mound.

Methane can smell bad as it burns, and for this reason the Barbers propose that the dragon in *Beowulf* is nothing more than a burst of methane escaping from a tomb that a torch-holding grave robber opened up and accidentally set ablaze. This is why the flames spouted by the dragon smell foul and why, after the dragon has "died," presumably because all of the methane has been consumed, Wiglaf can find no trace of the creature. There *was* no creature, just one imagined by those who could not comprehend flames leaping out of a burial mound all on their own.

It is worth noting, though, that while fire-breathing dragons are clearly present in the tales of Beowulf and the early Merlin stories, neither Tiamat nor Dracos, the creature in the story of the Golden Fleece, is ever described as having this ability. For that matter, fire-breathing dragons are not found in any Babylonian, Egyptian, Inca, or Native American mythology. There is a reasonable chance that this is linked to climate and geology.

It is well known that the Egyptians, Babylonians, Sumerians, Incas, and many other people built extensive tombs where they sealed away their dead, but methane explosions inside pyramids and other burial structures are unheard of. Part of the reason for this is related to the larger size of the tombs. Some Egyptian pyramids had vents running between different rooms, and this, along with their size, may have been enough to keep any methane that did get produced from accumulating. However, the arid climate where these people lived also played a role.

Dense and moist soils, which are common in northern Europe, allow almost no gases to pass through them. When used for building burial facilities, this soil creates the perfect gas trap. In contrast, dry and sandy soils, like those found in the Middle East and in the mountainous areas where the Inca dwelled, readily allow gases to pass through them. Under these conditions, even corpses buried behind a wall of soil inside a cave or pyramid would not have ended up in an oxygen-starved environment, since oxygen would constantly leak through any erected sediment wall. These geologic realities, along with the presence of extensive coal mines in Wales, may be why fire-breathing dragons seem to exist as monsters only in Europe.

But geology does not provide a perfect explanation for the phenomenon of fire breathing when Asia is taken into consideration. "Fire pit graves" have historically been described in the Yangtze region of China, where grave robbers foolish enough to carry torches as they broke into well-sealed tombs were met with fiery explosions. The geologic conditions in many regions of China were effectively very similar to those in northern Europe, and to make the situation even more dangerous, tombs built during the Han dynasty (c. 200 BC–c. AD 200) were often coated with thick layers of sticky clay. Robert Lee Thorp, a professor of Chinese art history and archaeology at Washington University in St. Louis, explains that the tombs easily trapped methane from decaying organic matter inside the chamber and that fires were often started by grave robbers who created sparks with their shovels. He also notes that Wang Ch'ung, a scholar who lived in China during this period, wrote that "flames shot out of a burial chamber and burned to death several hundred persons nearby."

A closer look at Wang Ch'ung's writings reveals the tale of two dead princesses. The myth describes the princesses as having had hatred in their hearts, and when people came to open the coffin of the first princess, a foul smell spread out and killed many. When the grave of the second princess was opened, the fire that incinerated hundreds blasted out. This hints that the malevolent spirits of the princesses themselves were somehow playing a part in these catastrophes, but it backs away from directly linking the events to the presence of monsters and definitely does not make a connection to dragons.

The Chinese did believe in dragons though. They even had a dragon in their stories known as Fu-ts'ang lung who functioned as a guardian of hoards, priceless jewels, and precious metals within the deepest and darkest vaults of Earth. Yet descriptions of this dragon breathing fire are not found in any art or literature. Part of the reason for this might be because Chinese dragons had a long history of not being viewed as monsters at all.

From dragon to deity

Today, good dragons, like the beasts in Rob Cohen's *DragonHeart*, Christopher Paolini's *Eragon*, and Dean DeBlois and Chris Sanders's *How to Train Your Dragon*, are as common as bad ones, but benevolent dragons are not a modern invention. The scholar Wang Fu, who was active during the Han dynasty, did not describe dragons as evil. Instead, he wrote about them as good and whimsical godlike creatures.

Like Zeus, Hera, and Apollo of the Greek pantheon, the Chinese dragons could fly, even if they were wingless; had magical powers, allowing them to control natural events (like the weather); and had tremendous shape-shifting capacities. They often brought rainfall to farmers whose lands were suffering droughts and luck to those in need. They were sometimes capricious in their ways, but mostly well-meaning.

Dragons were almost always associated with power, and numerous Chinese lineages claimed descent from unions that took place between mortals and dragons disguised in human form. The robes of the emperors were even adorned with the five-toed imperial dragon as a symbol of their status. However, unlike the gods of classical Greece, the dragons of China were clearly animalistic.

Wang Fu describes them as carrying the features of nine distinct entities, "The dragon's horns resemble those of a stag, his head that of a camel, his eyes those of a demon, his neck that of a snake, his belly that of a clam, his scales those of a carp, his claws those of an eagle, his soles those of a tiger, his ears those of a cow."* Dragons were obviously believed to be creatures of mixed qualities, and

*Amusingly, in the Disney film *Pete's Dragon*, as the little boy Pete describes the dragon he has become friends with, he sings, "He has the head of a camel, the neck of a crocodile, and the ears of a cow. He's both a fish and a mammal and I hope he'll never change." One has to wonder if Disney knowingly borrowed from Wang Fu. Perhaps a remake more faithful to history will one day be made that includes "the eyes of a demon" in the lyrics, but I doubt it.

paleontologically this makes sense. In the ancient Chinese text *I Ching,* which dates back to around 800 BC, it is noted that finding dragon bones in fields was viewed as an omen of good luck.* This work is clearly talking about fossils turning up in Chinese soil, but from Wang Fu's description, these "dragon bones" were not just the bones of ancient reptiles. China is loaded with other fossilized animals, including camels, cows, stags, clams, carp, etc. It seems that any fossilized bones, whether they were mammal, bird, reptile, fish, or invertebrate, simply became classified as having belonged to dragons.

The mixed traits were not just physical, though; ethically, dragons were also somewhat chimeric. While demons are bad, and tigers, eagles, and snakes are all predatory, camels, cows, stags, clams, and carp are all benign animals that play important roles in human life as either service beasts or food. Moreover, Wang Fu wrote that dragons had exactly 117 scales. Of these scales, 81 were infused with yang energy, which is a positive energy associated with goodness, and 36 were infused with negative yin energy. So they were mostly filled with yang but carrying enough yin to lead them to sometimes behave badly.

With the tomb guardian Fu-ts'ang lung, such a mixed description makes sense. Here was a dragon that was a protector of the dead and their belongings fulfilling a noble role, but also the ruler of an environment that generated deadly earthquakes and was filled with poisonous and sometimes explosive gases. In a way, this dragon served a purpose that was much like that of Pele in Hawaii, and he was in good company.

The sky-dwelling dragon of Chinese myth, Shen-lung, fits equally well into this dual persona. Best known as the storm bringer, Shen-

---

*Adrienne Mayor points out in *The First Fossil Hunters* that this was probably true in more ways than one. Dragons' bones were valuable in Chinese medicine (and still are in some regions). Fossils literally were a form of cash crop. Once farmers realized they had them on their property, they went searching for bones so they could sell them off and boost income.

lung had a terrible roar that could shake the entire landscape. The roar of this beast clearly invokes the idea of thunder, which would have been understandably frightening to ancient people. However, just as the storms that produce these effects can be scary and sometimes damaging, they are vital to human life. Without rain, there would be famine, and for this reason the dragon was a being embodying the two sides of life, good and evil, life and death.

To a certain extent, good, or at least neutral, dragonlike creatures found their way into the myths of other cultures. The Moki Indians, who lived in what is today the state of Arizona, worshipped a serpent known as Baho-li-kong-ya. Drawings of this beast often resemble a rattlesnake, which is logical considering where the Moki were living. Baho-li-kong-ya even had horns on its head just like the ancient drawings of Tiamat, but what is fascinating about this Moki creature is that it was also a fertility symbol. There are many snakes and snakelike beings in Egyptian mythology that also were not evil.

In light of the inherent fear that humans have of reptiles, a fear that often transformed these creatures into monsters, it is odd for them to attain a status in some cultures as deities embodying positive aspects of human life. Historians have tried over the years to make some sense of this. One theory centers around the fact that snakes shed their skins as they grow. Of course, we shed our skin too, but our skin cells do not all fall off at once revealing an entirely youthful-looking human underneath. For snakes, this is exactly what happens. Could this have led people to view snakes as somehow ageless and immortal? The idea certainly seems reasonable enough, but there is another possibility.

Snakes are cold-blooded, meaning their bodies are almost always the same temperature as their surroundings. If it is cold, they have difficulty moving because their chilled muscles function sluggishly. If it is extremely hot, they must seek shade or risk overheating. These realities make being warm-blooded, as mammals and birds are, seem like a blessing, but warm-bloodedness comes at an enormous cost. To maintain a constant body temperature, mammals and birds must constantly burn calories, and this means frequent feeding. Humans have it relatively easy because the diet for most people includes food that is packed with nutrients. Animals like cows and horses, which feed on nutrient-poor foods like grasses, must spend most of their days eating to get the calories they need. Reptiles gain a big benefit from their cold-bloodedness by not having to eat very often at all. Indeed, after a crocodile or snake makes a large kill, it may not need to seek out food again for several months. Because of this, ancient people may have viewed reptiles as somewhat supernatural in their ability to survive without food, raising them up from the role of monster to the role of deity.

Scaled back?

At first glance, dragons today seem to qualify as monsters just as they once did, but in large part, such a perception is an illusion. The dragons in the Harry Potter films are mystical beasts that certainly try to harm the protagonists, but they are not the stuff of nightmares. The situation is similar for the dragons found in the films *How to Train Your Dragon, DragonHeart,* and *Eragon*. They are impressive creatures, but audiences do not walk away fearful of dragons coming to eat them. Dragons no longer have the power to invoke fear as they once did.

The fact that dragons are not used as monsters is logical. Audiences know there are no dragons "out there," and this single fact makes establishing believability, which is crucial for fear to be created, very difficult. Think about it, there are no films that believably present dragons as being alive and posing a threat to people. This is not to say that no films have tried. Rob Bowman's 2002 film *Reign of Fire* made a valiant attempt at presenting dragons as real and frightening, but they never came across as a viable threat that audiences walked away scared of.*

---

*Consider *Jaws* as a comparison. How many people walked away from that film and nearly wet themselves at the idea of going to the beach or taking a sailing holiday?

But dragons are not entirely gone. To a certain extent, the presence of dangerous reptiles in modern films is a continuation of dragons as monsters just as it is a continuation of Medusa. *Lake Placid* and *Anaconda* are, at least partially, modern manifestations of dragons like Tiamat. However, dragons also live on in another form.

In films about fires, like Ron Howard's *Backdraft,* where the heroes race to fight numerous infernos set by a lunatic arsonist, the blazes function like different living creatures. Some creep, some fly, some are relatively benign, others violent. All play upon a seemingly deep-seated fear of fire, but none does this more effectively than the backdraft.

Backdrafts form when a large fire takes place in a contained space. At first the fire burns fiercely, but as oxygen in the space is consumed, the flames are forced to calm down even though plenty of flammable material remains. The fire is not dead. Quite the contrary, it is very much alive, just waiting for a breath of fresh air to allow it to explode back into a furious holocaust.

In the coal mine and tomb fires that likely led people to believe dragons could breathe flame, backdrafts would not have often (or ever) been involved. Even so, both situations led to the sudden appearance of a fiery ball of death—definitely something worth being scared of.

At the end of David Yates's *Harry Potter and the Deathly Hallows, Part* 2, Harry saves Draco Malfoy from a raging fire inside Hogwarts by grabbing him just as he is about to fall and burn. But this is not an ordinary blaze; it is shown with serpentine traits. A snake's head initially comes lunging out of the inferno, and at one moment the monster opens up bright orange wings of flame as it takes to the air. A nod to the origins of dragons?

6

Hauntings—Demons, Ghosts, Spirits

"Are you the Keymaster?"

—Dana Barrett/Zuul, *Ghostbusters*

Without a doubt, one of the most disturbing paintings of the 1700s is by the Swiss artist Johann Füssli. The work depicts a beautiful woman lying asleep on a bed. Sitting upon her chest is a demonic creature, staring at her with wicked intent. It was a day and age when monster movies and horror films didn't exist, but with a work of art as unsettling as *The Nightmare*, who needs Hollywood? The painting tells a story, and a creepy one at that.

The woman in the painting is widely thought to have been Füssli's beloved Anna Landolt. He and Landolt wanted to marry, but her parents had refused to allow it. Without the ability to sleep in her lover's arms, Landolt's lustful feelings draw the attention of a demon that comes to feed on her sexual energy as she rests. Some art historians argue that the painting is, in effect, a warning to parents, presenting the dangers of creating an impediment to love. Others suggest it is simply making a more general argument that dangerous things can come as the result of dreams. Either way, *The Nightmare* makes one thing very clear: A fear of dreams was present in Europe during the 1700s and it manifested itself in the shape of demons.*

The Nightmare, by Johann Heinrich Füssli. Oil on canvas, 1781. Detroit Institute of Fine Arts/Marc Charmet.

This is very different from the way things once were. The origin of the word "demon" is the ancient Greek word *daïmôn,* which comes from the verb *daïein,* meaning "to distribute." It was sometimes used during the days of Homer as a synonym for "god" but, more often than not, was associated with entities that distributed people's destinies. Unlike the gods on Mount Olympus, *daïmônes* were not actively worshipped. They were just poorly defined supernatural beings that

*The horse in the background is thought to be a visual pun on the word "nightmare" and was not part of the original chalk sketch of the work.

were neither really good nor really bad. Hesiod describes them as the watchers of Zeus who spy on mortals and as assistants to the gods. He also suggests that some people became *daïmônes* when they died. Plato theorizes that they were intermediaries between gods and men that were not gods themselves. They were sometimes associated with bringing dreams to people but were not specifically responsible for nightmares.

The appearance of Satan in mythology changed the role of *daïmônes* such that they started being viewed in a more negative way. Satan, or Ha-Satan, as he is known in the Hebrew Bible, is initially tasked by God to put the righteous believer Job through hardship to test whether he will prove faithful only during times of health and prosperity or at all times. Ha-Satan makes Job suffer terribly, but all of this suffering is brought about by Ha-Satan's powers alone; there are no servants working for him. However, with the rise of Christianity, Ha-Satan ceased to be an angel who tested the faith of humanity under the instruction of God and transformed into Satan, an enemy of God who commanded legions of twisted minions.

The once neutral *daïmônes* of the world came to be known as demons and started to function as an army of darkness that Satan could command to corrupt mortals in his war. People who were thought to be working with Satan came to be known as witches and wizards, and it was widely believed demons were sent to aid these villainous individuals in their work. These assistant demons, or imps, as they were called, often took mundane animal forms. They could appear as toads, rats, owls, goats, and cats so they could help witches and wizards without being too easily noticed by neighbors.*

By the late Middle Ages, demonology and the myths surrounding it were extremely rich and fears were surging. Like animals, demons of different sorts were treated as species that specialized in tormenting people in various ways, and those that came to be among the best

*This is where the concept of wizards having familiars comes from and why Harry Potter has an owl, Hermione a cat, Ron a rat, and Neville a toad.

known were the demons fueling feelings of lust: the succubus and the incubus.

The succubus was a she-demon intent on bending men to her will and siphoning away their souls through sex. The incubus was her male counterpart, a monster capable of corrupting young women, like Landolt in *The Nightmare,* and led these ladies to lives of debauchery. Both names derive from the sexual positions that these demons were thought to take. Succubus stems from the Latin word *succubare,* meaning "to lie under." Incubus stems from the Latin *incubare* for "to lie on top."

However, while *The Nightmare* makes it obvious that sexual demons were well known during the 1700s, they had actually been around for hundreds of years. A quick return to Geoffrey of Monmouth's *Kings of Britain* and Merlin's first meeting with the king Vortigern reveals the presence of sexual demons. After Merlin is collected by Vortigern's scouts, but before he makes his prophecy about dragons dwelling underground, his mother is grilled about how exactly she came to give birth to a child without a father. She responds:

> As my soul liveth and thine, O my lord the King, none know I that was his father. One thing only I know, that on a time whenas I and the damsels that were about my person were in our chambers, one appeared unto me in the shape of a right comely youth and embracing me full straitly in his arms did kiss me, and after that he had abided with me some little time did as suddenly vanish away so that nought more did I see of him. Natheless, many a time and oft did he speak unto me when that I was sitting alone, albeit that never once did I catch sight of him. But after that he had thus haunted me of a long time I did conceive and bear a child.

Similar to a claim of immaculate conception, Merlin's mother argues that a phantom lover kissed her, spoke to her, and eventually impregnated her. Vortigern is shocked by this, calls for an adviser named Maugantius, and asks if he believes Merlin's mother's story. Maugantius responds: "In the books of our wise men and in many histories have I found that many men have been born into the world on this wise. For, as Apuleius in writing as touching the god of Socrates doth make report, certain spirits there be betwixt the moon and the earth, the which we do call incubus daemons." And Geoffrey of Monmouth wrote more than 500 years before *The Nightmare* was painted.

Seven hundred years before Geoffrey, the evolution of sexual demons from more neutral *daïmônes* is seen in St. Augustine's *De Civitate Dei* (*City of God*): "It is a widespread belief that sylvans and fauns [nature spirits], commonly called incubi, have frequently molested women, sought and obtained coitus from them." St. Augustine's writings are about the earliest that we find mentioning the threats presented by the succubus and incubus, but the odd thing about this medieval rise of sexual demons is that they seem to represent a fear of sexual seduction that could hardly have been new.

Siren song

In the *Odyssey,* Odysseus runs into numerous temptations and threats of seduction. The Lotus-Eaters try to lure him into a life of eternal flower eating, and the beautiful witch Circe does her best to convince Odysseus his place is to be forever by her side, but these threats are nothing compared to that presented by the Sirens. As Odysseus prepares to leave Circe, she sternly warns him, "First you will come to the Sirens, who enchant every single man who comes to them. If anyone draws near to them in ignorance and hears their voices, there is no homecoming . . . instead he is enchanted by the clear, sweet song of the Sirens who sit in a meadow, surrounded by a great heap of rotting men, skeletons with shreds of shrivelling skin on them."

Remarkably, when Odysseus reaches the Sirens, the text gives absolutely no physical description of these monsters while describing their song in considerable detail: "Come here, illustrious Odysseus, great glory of Greece, beach your ship, so you can listen to our voices. For nobody has ever sailed by on his black ship without listening to the honeyed words on our lips."

Depictions of Sirens in art and literature made many years after Homer show these monsters as bird-women, presumably because birds sing and because many birds are, in fact, carnivorous and sit in nests surrounded by skeletons and rotting flesh. Yet Homer's decision to not describe the Sirens' bodies is worth noting, since so many other monsters in his stories are described in great detail.

Intriguingly, even the gender of the Sirens is concealed. They have honeyed words on their lips and they are specifically mentioned by Circe as attracting men, so they are often assumed to have been female, but Homer does not actually state that they are. One reason for the gender and physical ambiguity in the *Odyssey* could be that the Sirens were so well known that everyone hearing the story was expected to know what they looked like. Given the extensive descriptions of so many other monsters, however, it seems more likely that Homer simply wanted to have a monster in his story that represented the fears associated with the temptations of the flesh. By presenting the Sirens without physical form, he effectively left which temptations they represented up to the listener.

There are obvious similarities between these Greek monsters and the demons that appear so much later in history. All of them seduce, but there is a key difference between the Sirens and the demons found later in *Kings of Britain* and *The Nightmare*—sleep.

Sweet dreams

People who were forced to face the Sirens did so awake. If they fell prey to them, it was because of their own conscious folly. In *The Nightmare,* it is sleep that is the distinctly new element mixed with sexual seduction to create the monster. In *Kings of Britain,* the situation for Merlin's mother is less obvious. The text hints that she does ultimately sleep with the demon, since it leads her to conceive a child. Moreover, it is described as coming to her "chambers," suggesting these were nocturnal visits since time spent in chambers would most often be at night when she was getting ready for bed or sleeping. Even so, it is not entirely clear if her interactions with the demon took place when she was awake or dreaming. Yet there is a progression present that suggests that the evolution of the incubus demon from *daïmônes* and monsters like the Sirens was not anxiety over sex but rather a rising fear of sexual thoughts taking place during sleep. This is strange because studies of ancient human communities hint that, historically, people have had no problem with sexual thoughts during sleep.

The Hadza in northern Tanzania have had limited exposure to the ways of the outside world, and their traditions are much the same today as they were thousands of years ago. They don't mix with other tribes, rarely paid attention to Europeans who tried to make contact, and their language, which involves a number of click sounds, is a challenge for Westerners to learn. They are thus rather well isolated from many of the ideas floating around in the rest of society.

As for how they handle the matter of sexual dreams, the Hadza have treated a girl's first menstruation as a joyous occasion for generations. The girl is adorned with beads and a celebration is thrown in her honor. Intriguingly, when Hadza boys ejaculate in their sleep for the first time (an event that is often associated with dreams of sexual activity) they too are adorned with beads and given a celebration. The sexual dream is not something to be ashamed of but something joyous.

Even so, the Hadza are not a perfect window into the past. Some of their traditions may have evolved with time, and their treatment of girls and boys as they make their journey into adulthood may have altered somewhat from what it was a few thousand years ago. However, it is hard to imagine the celebratory behaviors that anthropologists see in the Hadza today stemming from behaviors that were once associated with any sense of shame. Minor changes would be understandable, but a dramatic change in overall tone seems unlikely.

Other relatively isolated tribes, like the Umeda of Papua New Guinea, show similar tendencies. It is not uncommon for hunters, on the night before a major hunt, to sleep on top of specially scented sacks that lead them to have erotic dreams and orgasms. Experiencing nocturnal ejaculations was considered to be a good omen, since sex between hunters and women usually followed a successful hunt and dreams of sex hinted that sex would come for real in the near future.*

So if the Hadza and Umeda can be seen as models of how people have historically responded to sexual dreams, this suggests that the ancient human condition embraced rather than feared sexual dreams. Somewhere along the way, during the evolution of society, the wheels came off the wagon and sexual dreams became something to dread. Can we blame the Greeks or Romans for this?

In ancient Greece, sexual dreams appear to have also been a cause for joy. It was common for dreams to be interpreted, and Herodotus wrote about the Greek traitor Hippias, who helped the Persians during their invasion when they landed at Marathon. Hippias had a dream of sleeping with his mother: "He dreamt of lying in his mother's arms, and conjectured the dream to mean that he would be restored to Athens, recover the power which he had lost, and afterwards live to a good old age in his native country. Such was the sense in which he interpreted the vision." This led him to believe he would one day return to his motherland of Athens in a position of power.

Similarly, Artemidorus, a Roman dream interpreter who lived in the third century AD, wrote that having sexual dreams involving one's mother were good for people working in politics, since such dreams represented the love one felt for one's motherland and were a sign of deep-seated patriotism. Yet Artemidorus was doing his work at a time when sexual dream interpretation was changing, and according to the anthropologist Charles Stewart at University College London, who wrote extensively on nightmares in the *Journal of the Royal Anthropological Institute* in 2002, incubi in very early forms started to leak into his writings.

---

*This is actually remarkably similar to behavior seen in numerous primates where there is a dramatic surge in sexual activity shortly after securing a major food source.

In his *Interpretation of Dreams*, Artemidorus makes a connection between the playful god Pan, who was known for engaging in dream mischief, and an entity who is mysteriously named Ephialtes. According to Stewart, Ephialtes etymologically seems to mean "to jump on top of," which is not much different from "to lie on top." More specifically, he points out that Artemidorus writes, "Ephialtes is identified with Pan but he has a different meaning. If he oppresses or weighs a man down without speaking, it signifies tribulations and distress." Although the word "incubus" is not specifically used, there are some uncanny similarities between Ephialtes and the monster appearing in Füssli's *The Nightmare*.

Old Man and a Siren. Roman marble relief, fragment, second century AD. Museum of Fine Arts, Boston. Bridgeman Art Library.

Artwork supports the idea that AD 200 was a time when sleep demons were beginning to become more widely feared. In a marble relief at the Museum of Fine Arts in Boston, a sleeping shepherd is straddled by a voluptuous winged woman with webbed feet. Unfortunately, the relief is fragmentary and the head of the woman has been lost to the ravages of time. For this reason we cannot determine how human she actually is or get a sense of whether she wishes the sleeping shepherd sweet dreams or sexual nightmares. Even so, many argue that the woman is, in effect, a Roman representation of one of Homer's Sirens. This has led to the relief being named *Old Man and a Siren.*

The timing of this transformation of Sirens into sleep demons, like the incubus and succubus, almost perfectly aligns with when Christianity was spreading rapidly throughout the classical world. As Christians came to blame many misfortunes on evil spirits, the neutral *daïmônes* of ancient Greek lore were transforming into Satan's fiendish minions. When people ate too much, they fell prey to the demon of gluttony. If they didn't do enough work, they were plagued by the demon of sloth.

Demons associated with behaviors that people could control with effort were frightening, but demons of nocturnal sexual thoughts must have been dreadful because (in spite of what many believed) there was nothing anyone could do to control his or her dreams. By identifying dreams as potentially sinful, Christianity effectively led people to fear what their own minds subconsciously conjured. Upon waking, they would be forced to struggle with powerful feelings of guilt, corruption, and the terror of eternal damnation. Both the dreamers themselves and the clergy whom they sometimes confessed to could not or would not accept that such dreams were naturally generated by the minds of good people all on their own. Something, or some things, had to be responsible. Thus the incubus and succubus evolved from their less overtly sexual relatives the Sirens and the less malevolent *daïmônes.*

Lingering in limbo

Even if Christianity was responsible for the transformation of sirens and *daïmônes* into sleep demons, there are still elements of their evolution as monsters that raise questions. The most perplexing of these is the strikingly different forms they take. In the marble relief *Old Man and a Siren,* the monster is standing over the man. In *Kings of Britain,* the demon merely visits Merlin's mother at night. In contrast, Artemidorus describes Ephialtes as a creature that is "jumping on top" and "weighing a man down," and in *The Nightmare,* the demon is literally perched on the center of the woman's chest.

If only a fear of sexual dreams was responsible for the formation of sleep demons, one would expect their forms to be reasonably consistent, but this is not what we find. Some sleep demons tend to be associated with chest pressure while others only suggest sexual arousal. This could be because people were trying to come to grips with more than just dreams of sex.

As the body enters the stage of sleep when dreaming takes place, known as rapid eye movement or REM sleep, the brain effectively initiates a safety mechanism that tells almost all muscles in the body to stop acting. For example, if a dream emerges that leads you to believe you are walking, you do not literally start walking while in bed. The same is true with more violent activities. If you dream about punching someone in the nose, your body, because of the safety mechanism, does not act upon this impulse.* Yet the safety signal that the brain sends out does not always work as well as it should.

In some situations, for reasons that are not entirely understood, muscles still take action even though they have been commanded by the brain to remain motionless. Under such a condition, known as REM sleep behavior disorder, people act out their dreams physically. This is why sleepwalking and, in a few noteworthy cases, even sleep-

---

*A good thing for people whose partners have particularly vivid imaginations.

driving, take place. This obviously dangerous disorder needs close monitoring.

A reverse condition, which is nowhere near as hazardous but often far more frightening, is sleep atonia, or sleep paralysis. In this situation, the safety signal works too well. When the person wakes, his brain ends its dream state and he becomes aware of reality, but his muscles continue to obey the "don't move" command that they were sent when REM sleep began. As a result, the waking person cannot move. He is literally frozen in place.

Sleep atonia sounds like the sort of thing that would be rare, but the medical literature indicates the opposite. According to the American Sleep Disorder Association's *Diagnostic and Coding Manual,** between 40 and 50 percent of the human population experiences sleep atonia at some point. More important, the manual explains that the condition is frequently associated with hallucinations.

As the mind shifts between the dream state and the waking state, it is common for it to notice things that aren't actually there. Sometimes noises are heard that belong to a dream rather than reality; sometimes the presence of other people, who existed in the dream, are sensed by the waking dreamer. These are hallucinations, and they are entirely normal and benign for most people, but when combined with sleep atonia, they can become truly awful.

One particularly frequent hallucination that seems to run hand in hand with sleep atonia is believing a creature of some sort is present in the room with the waking dreamer. Commonly, this entity is described as seated on the waking dreamer's chest, holding them in place or standing over them. In the Western world, this combination of hallucinations and sleep atonia takes place often enough for it to have a name: Old Hag Syndrome.

The name stems from an area in northeastern Newfoundland where a mix of hallucinations and sleep atonia appears to be remark-

---

*Suffer from insomnia? Try reading the ASDA *Diagnostic and Coding Manual*; it'll knock you out faster than Ambien.

ably prevalent, and people often refer to the hag or *ag* coming to terrorize them. According to the medical sociologist Robert Ness at Augusta State University in Georgia who reported on the condition in a Newfoundland community during the 1970s, people in the region commonly believed that the traumatizing sleep experiences were the result of being cursed by an old hag who came to them during the night to sit on their chest.

Observations of this condition are not unique to the English-speaking world. The Chinese tell tales of the *gui ya shen,* which literally means "phantom that presses on the body"; the islanders of St. Lucia in the Caribbean tell of the *kokma,* which is the spirit of a dead baby that haunts an area and attacks people in their beds by jumping on their chests and clutching at their throats; and the people of Thailand speak of the *Phi um,* an enveloping ghost that holds people in their beds as they wake up. When these legends are considered in combination with Artemidorus's description of Ephialtes jumping on chests, it seems that fear of sleep atonia has been present in communities around the world for centuries and long taken the form of something spectral dwelling half in the waking world and half in a world of dreams.

Evil evolved

So it would seem that at some point around AD 200, the terror presented by sleep atonia started to merge with the fear of sexual dreams induced by Christianity. The two fears became one and ultimately developed into the incubus presented in *The Nightmare.* And the fear has not gone away.

Modern films like James Watkins's *The Woman in Black* (which is an equally terrifying stage play) and Oren Peli's *Paranormal Activity* and television series like *The X-Files* gain much of their fear factor from the concept of creatures lurking in the realm between the seen and the unseen. Even though these creatures can be playful, like the succubus that comes to the heroes as they are sleeping in Ivan Reitman's comedy *Ghostbusters,*[*] more often than not they are truly terrifying, like the demons in Tobe Hooper's thriller *Poltergeist,* which upon emerging from the family's television caused the child Carol Anne to utter the immortal words "They're here."

An intriguing element of many stories about ghosts and demons is that they usually take place in hot spots for spectral activity. This is certainly the case in *Ghostbusters,* where Egon Spengler says, "It's not the girl, Peter, it's the building," when he defends the lead character's love interest, Dana Barrett, who is possessed not because she is inherently evil but because the building she lives in was built by a demon-worshipping lunatic. This is equally true of Mary Lambert's version of Stephen King's novel *Pet Sematary,* where there is much talk of "sour" land; *Poltergeist*; Stanley Kubrick's presentation of King's *The Shining*; and countless other phantasmal thrillers. Haunted places have a long history, and this leads to the natural question of whether there are certain environmental conditions, which could be present in a specific location, that can induce a combination of sleep atonia and hallucinations. A search through the dream disorder literature indicates there definitely are conditions, like jet lag, alcohol abuse, depression, and chronic anxiety, that can increase a person's chances of having a sleep atonia episode. However, none of these can reasonably be tethered to a location. So it would seem that there must be something else that makes locations haunted.

While most people suffer sleep atonia only once during their lives, a very small percentage of the population suffers regularly from the dreaded experience. Most important, this chronic version of the disorder is genetic; meaning parents who have it can pass it along to

*The title of this film is a terrible misnomer. The demon Zuul is the creature that possesses the body of Dana Barrett; the doglike demon Vinz Clortho takes over the body of Dana's neighbor Lewis Tully; and Gozer is the shape-shifting demon that initially appears as a red-eyed human and later transforms into the Stay Puft Marshmallow Man that the heroes do battle with during the climax of the film. While it probably would have driven away audience members by the thousands, *Demonbusters* would have been a more accurate title.

their children. This inherited version of sleep atonia may not seem especially relevant today, when children break away from their parents as young adults and tend to live in distant locations, but historically it was common for houses to pass from one family member to another over the generations. For this reason, it is possible that the concept of the haunted house emerged as a result of a family with the genes for chronic sleep atonia moving in and passing the home along to other family members who also carried the genes for chronic sleep atonia. Ironically, if this were true, it was not houses that were cursed but the families dwelling in them who were cursed with bad genes.

All hauntings and modern ghost stories aside, there are few recent monsters that are more true to the demonic history of mixing fears of uncontrolled sleep behaviors and sexual activity than Freddy Krueger from the 1984 film *Nightmare on Elm Street*.

In Wes Craven's film, a group of sexually inquisitive teenagers begin experiencing terrible dreams. They find themselves in a boiler room being chased by a disfigured man whose right hand is covered in a glove with wicked blades attached to every finger. Dreams and reality merge as the teenagers begin to find five-fingered blade marks in their clothes and sheets when they wake up, hinting that they could die from their dreams. This ultimately happens when one girl, Tina, is slashed to death in the night after sleeping with her young boyfriend.

Tina's murder, while entirely spectral in nature, carries sexual overtones. Later in the film, when Krueger's bladed glove appears in the bathtub between another teenage girl's legs, it is hard to not think of some sort of deadly phallus. The horrifying combination of the vulnerability of sleep and sexual violation links Krueger to the fears of the incubus.

Yet the film backs away from presenting Freddy Krueger as anything more than a deformed killer. Even after being disfigured by fire, he is still recognizable as having once been human. This is likely the result of the film's creators not wanting to stray too far from reality. If Krueger had been an utterly inhuman demon, he would have been less real, less believable, and, in the end, less scary.

And if Krueger is an evolved form of the incubus, so too are the many aliens that "abduct" humans. Charles Stewart convincingly argues that people who claim to have been abducted and sexually assaulted by aliens during their sleep are really just doing their best to rationalize experiences of sleep atonia, hallucinations, and sexual dreams. This makes perfect sense, particularly if people suffering from these sleep disorders no longer find demons believable but do consider the existence of extraterrestrials a possibility. The monster is being created by the same core fear, but believability is forcing the form of the monster to change.

7

Cursed by a Bite—Vampires, Zombies, Werewolves

"You're intoxicated by my very presence."

—Edward Cullen, *Twilight*

Slinking through the shadows of night, they come to feed on the innocent. Seemingly human in appearance, the threat that they pose becomes apparent only as needle-sharp fangs pierce the throat of their intended victim and blood is sucked away. When every last drop of this precious life essence is consumed, prey becomes predator, seeking out blood to fuel its own newly acquired supernatural hunger. Vampires are among the world's most celebrated and popular monsters, and they have an extremely complex history and biology surrounding them, supported by a long line of books and movies featuring them as both villains and heroes. Yet working out exactly which fears drove the rise of vampires is a tricky question to answer because they are such multifaceted monsters with no clear point of origin.

On the face of it, they are predators like lions and play upon the terror of being killed by a nocturnal hunter. With such a basic fear, one would expect vampires to be present during ancient times when fears of beasts lurking in the night were at their height, yet vampires as we know them today arrived on the scene only in the eighteenth century. Even so, earlier reports of creatures resembling these monsters do exist.

In the *Odyssey,* Odysseus is forced to travel to the land of the dead and confront the ghosts of people he once knew in order to gain information to aid him on his quest. The witch Circe advises that he must allow the ghosts to feed on blood freshly spilled from the body of an animal to gain their trust and knowledge. At first he is highly protective of the pool of blood that he spills on the ground, allowing only the ghost of the wise man, Teiresias, to feed and answer his questions. But then the ghost of Odysseus's mother appears and fails to recognize him as her own son. Odysseus turns to Teiresias for answers: "Tell me and tell me true, I see my poor mother's ghost close by us; she is sitting by the blood without saying a word, and though I am her own son she does not remember me and speak to me; tell me, Sir, how I can make her know me." Teiresias replies, "Any ghost that you let taste of the blood will talk with you like a reasonable being, but if you do not let them have any blood they will go away again." Odysseus then allows the ghost of his mother to feed on the blood, and her memories of him come flooding back.

For Homer, blood is clearly a link between the dead and the living, even if it has to be spilled from an animal's body onto the ground to have this effect. However, while the spirits in the *Odyssey* are a tantalizing ancestor to the modern vampire, they are still very different, and it is not until nearly two thousand years after Homer, during the late 1100s, that creatures more like the vampires of modern fiction appear in Europe. The person who documents these monsters is William of Newburgh, an English historian who is widely thought to have had a network of trustworthy informants who helped him report on historic events that took place between the days of William the Conqueror in 1066 and those of Richard the Lionheart in 1198. In his *Historia rerum Anglicarum,* amid information about royalty and political events, he tells the tale of an evil man who dies from a fall shortly after discovering his wife is having an affair:

> A Christian burial, indeed, he received, though unworthy of it; but it did not much benefit him: for issuing, by the handiwork of Satan, from his grave at night-time, and pursued by a pack of dogs with horrible barkings, he wandered through the courts and around the houses while all men made fast their doors, and did not dare to go abroad on any errand whatever from the beginning of the night until the sunrise, for fear of meeting and being beaten black and blue by this vagrant monster. But those precautions were of no avail; for the atmosphere, poisoned by the vagaries of this foul carcass, filled every house with disease and death by its pestiferous breath.

Death comes to many people after the evil man's burial, and desperate to bring the monster's curse to an end, two brothers take action:

> Two young men, who had lost their father by this plague, mutually encouraging one another, said, "This monster has already destroyed our father, and will speedily destroy us also, unless we take steps to prevent it. Let us, therefore, do some bold action which will at once ensure our own safety and revenge our father's death. There is no one to hinder us; for in the priest's house a feast is in progress, and the whole town is as silent as if deserted. Let us dig up this baneful pest, and burn it with fire." Thereupon snatching up a spade of but indifferent sharpness of edge, and hastening to the cemetery, they began to dig; and whilst they were thinking that they would have to dig to a greater depth, they suddenly, before much of the earth had been removed, laid bare the corpse, swollen to an enormous corpulence, with its countenance beyond measure

> turgid and suffused with blood; while the napkin in which it had been wrapped appeared nearly torn to pieces. The young men, however, spurred on by wrath, feared not, and inflicted a wound upon the senseless carcass, out of which incontinently flowed such a stream of blood, that it might have been taken for a leech filled with the blood of many persons. Then, dragging it beyond the village, they speedily constructed a funeral pile; and upon one of them saying that the pestilential body would not burn unless its heart were torn out, the other laid open its side by repeated blows of the blunted spade, and, thrusting in his hand, dragged out the accursed heart. This being torn piecemeal, and the body now consigned to the flames, it was announced to the guests what was going on, who, running thither, enabled themselves to testify henceforth to the circumstances. When that infernal hell-hound had thus been destroyed, the pestilence which was rife among the people ceased, as if the air, which had been corrupted by the contagious motions of the dreadful corpse, were already purified by the fire which had consumed it.

The monster is never called "vampire," but the connections to the creatures we know as vampires today are strong. He was "issuing, by the handiwork of Satan, from his grave at night-time" and responsible for the deaths of many people in the surrounding area. Then, when the brothers attack him, William states that they "inflicted a wound upon the senseless carcass, out of which incontinently flowed such a stream of blood, that it might have been taken for a leech filled with the blood of many persons." A leech is a bloodsucking creature, but this passage steers clear of actually saying this was a bloodsucking monster.

However, a return to the original Latin hints that there is more: "*Nec territi juvenes, quos ira stimulabat, vulnus exanimi corpori intulerunt: ex quo tantus continuo sanguis effluxit ut intelligeretur sanguisuga fuisse multorum.*" This can be translated as, "The brave young men, excited by wrath, struck a wound on the lifeless corpse, from which so much blood then flowed that it was understood that he had been the bloodsucker of many."*

Like Homer's ghosts, we again find a connection between the undead and a thirst for blood. The difference this time is that the undead creature in this story is definitely malevolent and bringing harm to people in the real world, whereas the ghosts in the *Odyssey* are not harming anyone.

William of Newburgh told many similar stories of the dead rising from the grave, and he had much company. In 1591, in the town of Breslau (now the Polish city of Wrocław), a shoemaker who killed himself by putting a knife through his neck, came back to haunt those around him by pressing against their necks in the night. He was ultimately found in his grave with the wound in his neck just as fresh and red as it had been when he died. In 1746, the French abbot Augustin Calmet reported, "A new scene is offered to our eyes. People who have been dead for several years, or at least several months, have been seen to return, to talk, to walk, to infest the villages, to maltreat people and animals, to suck the blood of their close ones, making them become ill and eventually die."

The solution to the undead threat that locals turned to was exactly what William of Newburgh described. They dug up the graves of the offending monsters to destroy them and found that recently buried corpses often had blood on their lips, bloated stomachs that looked as if they had just fed, blood still flowing inside their bodies, fresh-looking organs, clawlike fingernails, and elongated canine teeth. Terrified by these sights, people chopped off heads, drove stakes through hearts, and jammed bricks into decaying mouths to keep the monsters from biting anything more. It must have been dreadful business, but there are no reports of the mon-

---

*The Latin word *sanguisuga* literally means "blood" (*sanguis*) "sucker" (*suga*). It is understandable that the original translation used the word "leech" in place of "bloodsucker" since "bloodsucker" is not really a word typically thrown around in scholarly English . . . unless, of course, you happen to be writing a book about monsters.

sters ever fighting back. They are always just corpses in graves taking a beating.

Finally, after hundreds of years of terrorizing Europe, all these walking corpses and ghosts earn the name "vampire" in the second edition of the *Oxford English Dictionary* in 1745. It was described as "a preternatural being of a malignant nature (in the original and usual form of the belief, a reanimated corpse), supposed to seek nourishment, or do harm, by sucking the blood of sleeping persons." People must have been scared out of their socks.

Mortifying misunderstanding

With so many traits and behaviors being associated with these early vampires, it is likely there were several fears merging together to form these monsters. As such, it seems best to start with the most concrete details being described: Vampires had bloody mouths, bloated stomachs, fresh blood in their bodies, and, sometimes, claws and fangs.

The Europeans who were initially digging up corpses were probably not exaggerating. After people die, bacteria living within the body often continue to be productive and generate gases that collect inside. The gas production leads to an effect that morticians refer to as "postmortem bloat," and while it has nothing to do with diet or recent feeding, it can make the belly look swollen and lead people to conclude that the corpse has recently eaten.

In addition, gas buildup inside the body can cause blood to get pushed up from the lungs, passed through the trachea, and out of the mouth so that it stains the teeth and lips. This likely created the illusion that the bloated stomach was not simply full, but full of blood that the corpse had recently consumed, logically leading to the idea that the monster fed on blood.

Furthering the idea of the animated corpse, under certain circumstances bacteria-created gases can move past the vocal cords and create sound. This often occurs when bodies are handled or meddled with after death, causing corpses to make noises as if they are groaning or, in rare cases, speaking.*

As for elongated canines and clawlike fingernails, there is a medical explanation for this too. After death, tissues die and waste away; the skin begins to shrink, and this leads it to be pulled back along both the nail beds and the gum line. As a result, the nails and teeth become more prominently exposed than they were at the time of burial. Of course, this is an illusion, but to early vampire hunters who had worked themselves into a lather over the perceived plague of the undead, these were fangs and claws indicative of a vampiric transformation.

All of these natural processes can explain the descriptions of early vampires and can even account for why Homer, way back in ancient Greece, suggested that the dead liked to feed on blood. But one thing that is not immediately clear is why the belief of the dead leaving their graves to attack the living gained such popularity during the 1100s when William of Newburgh was writing but not during the days of Homer. One possibility worth considering is that people being buried during William's time were not actually dead.

Today there are a lot of tools available, like blood pressure cuffs, stethoscopes, and heart monitors, that help doctors determine whether someone is alive or dead. Yet even with these devices, patients with very weak or infrequent heartbeats can easily be declared dead by mistake. As an example, in Jan Bondeson's book *Buried Alive,* which goes into great detail on how accidental burials happened (and still do), the tale is told of a Frenchman named Angelo Hays who suffered a brutal motorcycle accident in 1937. At the hospital he was not breathing, had no detectable pulse, and had a serious head injury. The doctor, using a stethoscope, could not hear anything, and Hays was sent to the morgue. Three days later, as he was buried, an insurance company realized Hays had been covered by a policy for up to 200,000 francs and sent an inspector out

*Driving a stake through a corpse's chest counts as meddling at the highest level.

to investigate the accident before paying up. The inspector ordered the body exhumed to look at the injuries and to confirm the cause of death. Remarkably, the doctor in charge found the corpse to still be warm.* Hays returned to the land of the living and is thought to have survived his near-death ordeal by being buried in loose soil that allowed some flow of oxygen to the coffin and by needing very little oxygen in the first place as the result of his head injury reducing all metabolic activities in the body. Bondeson relates a few more similar stories and argues that if we see such cases now, they probably were taking place somewhat more often in the past when vital sign monitoring tools were not available. Could such events of still-living but "geologically challenged" patients have been feeding into undead mythology?

Rising from the grave

In 1938, the author, folklorist, and anthropologist Zora Neale Hurston, then a student of the noted anthropologist Franz Boas at Columbia University, proposed there might be some material basis for the stories told in Haiti of individuals being raised from their graves by voodoo masters. These raised people, or zombies, legends said, were robbed of their identities, enslaved, and forced to work indefinitely on plantations. Hurston was not believed. For decades, the wider research community ignored her suggestions and in some cases actively ridiculed her, but this attitude eventually changed.

In May 1962, a man spitting up blood and sick with fever and body aches sought help at the Albert Schweitzer Hospital, a facility operated in Haiti by an American charity. Two doctors, one of whom was an American, did their best to save him, but to no avail. The

*"Buy our life insurance and we promise that our greed will ensure you are most certainly dead before we pay." Not exactly an advertisement that any company is likely to use today, but let's face it, they saved Hays's life.

man's condition deteriorated and he was declared dead shortly after his arrival. At the time of his death, he was diagnosed as suffering from critically low blood pressure, hypothermia, respiratory failure, and numerous digestive problems. What exactly caused such systemwide problems remained a mystery. The man's sister was called in to identify his body and stamped her thumbprint to the death certificate to confirm he was her brother and that he was, indeed, dead. Eight hours later he was buried in a small cemetery near his village, and ten days later a large stone memorial slab was laid over his grave.

In 1981, the sister was approached by a man at her village market who introduced himself to her using the boyhood name of the dead brother. It was a name that only she and a handful of other family members knew, so he seemed real enough. The man explained that he had been made into a zombie and forced to work on a sugar plantation with many other zombies until their master died and the zombies were freed. The media went crazy with the story, particularly in Haiti, and Lamarck Douyon, the director of the Psychiatric Institute in Port-au-Prince, made up his mind to test whether this zombie tale could possibly be true.

Douyon knew that digging up the grave would prove nothing; if the man and his zombie story were fraudulent, it would have been easy for the deceivers to remove remains from a rural village cemetery. Instead, Douyon collaborated with the family to construct the ultimate identity test. He would ask the man a series of questions that only the brother would know all the answers to. The man passed the test, and later, when the sister's thumbprint and the thumbprint on the death certificate were confirmed by Scotland Yard to be identical, Douyon concluded the man's story was likely true. There had to be something real about the zombie mythology of the island.

All of the evidence pointed to the idea that some sort of a poison had been used to make the man appear dead after making him quite ill. Then, after he was buried, he had been exhumed by his poisoner so he could be enslaved. Realizing that this was a matter for a biochemist rather than a psychologist, Douyon and other doctors in Haiti asked the Harvard ethnobiologist Edmund Wade Davis to get involved.

Davis conducted several expeditions to Haiti and collected five zombie poison recipes from four different locations. All the poisons varied in the number of tarantulas, lizards, millipedes, and nonvenomous snakes added to the brew, but there were a handful of similarities that caught Davis's attention. All recipes contained a species of ocean-dwelling worm (*Hermodice carunculata*), a specific tree frog (*Osteopilus dominicensis*), a certain toad (*Bufo marinus*), and one of several puffer fish (also known as blowfish in some regions).

Since these organisms appeared in all the different zombie poisons, Davis focused his attention on them. He found that the worm had bristles on its body that could paralyze people, and the tree frog was closely related to a frog species that released toxins on its body that could cause blindness in those who touched it. Furthermore, the toad, he learned, was a chemical nightmare. Some of the compounds in its body functioned as anesthetics, some as muscle relaxants, and some as hallucinogens. He noted that earlier studies conducted with the toad compounds had discovered they induced a rage similar to the berserker rages found in Norse legends, and these studies suggested that compounds of closely related toads had once been consumed by ancient barbarians as they charged with reckless courage into battle and shrugged off all but the most lethal attacks.* But by far the most interesting ingredients in the zombie poisons were the puffer fish.

Puffer fish, which are well known for their deadly nerve toxins, are said to be tasty. Eating them comes with the serious risk of being poisoned, but this doesn't put off the Japanese. Called fugu in Japan, puffer fish is something of a dining adventure that popularly leaves

*In some tales, these berserking warriors are said to have transformed into bears to maul their enemies. Whether the drugs they were taking led the warriors to view themselves as werebears or whether the sight of them charging in a bestial fury covered in animal furs led their foes to believe they had become animals is difficult to determine. Either way, it seems the toad was involved.

consumers with feelings of body warmth, euphoria, and mild numbness around the mouth. Of course, if the chef gets fugu preparation wrong, diners end up in the hospital. Because the fish is so popular, hospitalization occurs with relative frequency and, as a result, there is a lot of medical literature on what fugu poisoning looks like.

Common symptoms include malaise, dizziness, nausea, vomiting, very low blood pressure, headache, and initial numbness around the lips and mouth that spreads to the rest of the body and often becomes severe. Eyes become glassy, and patients who survive the experience say it felt as if their bodies were floating while they could not move. They remained fully aware of their surroundings and alert during the poisoning experience. In one dramatic account, a fourteen-year-old boy in Australia, who accidentally ate puffer fish while on a camping trip with his family, recalled his family talking in the car as he was taken to the hospital, the nurses wishing him good morning and good night, and the doctors speaking their medical mumblings all while entirely paralyzed and feeling "light."

Davis found this intriguing because when he interviewed the man who claimed to have been made a zombie, he learned that he had remained conscious the entire time, heard his sister weeping when she was told that he had died, and had the sensation of floating above the grave. These descriptions, in combination with the medical reports filed when the man had been in the hospital on the night of his "death," suggested that puffer fish poison had been at work.

Upon further investigation, Davis learned that zombie makers created their poisons and exposed victims to these toxins by releasing them in the air near where the person lived or by putting them in places where the person was likely to make contact, such as on door handles or window latches. Lacing food with the poison was never done, because zombie makers believed it would kill the victim too completely.

After burial, zombie makers had their assistants pull victims out of their graves and then beat them fiercely to drive their old spirit away. This was followed by binding the exhumed person to a cross, baptizing them with a new zombie name, and force-feeding them with a paste made from sweet potato, cane syrup, and *Datura stramonium,* one of the most hallucinogenic plants known.

Davis suggested in the *Journal of Ethnopharmacology* in 1983 that these ghastly experiences, combined with the potent initial poisoning, created a state of psychosis that literally transformed people into zombies who would do anything they were told by their masters. This, he argued, explained why voodoo magic was widely perceived as raising the dead and why Hurston was right. People literally were being buried alive and then dragged back to the living world as zombies.

Undead plague

But what of vampires? The early stories about these monsters do not support the "buried alive" theory very strongly, and there are no indications of poisons or zombie makers being involved with vampire creation. The vampire historian Paul Barber points out in his book *Vampires, Burial, and Death* that none of the early vampiric accounts actually describe vampires digging themselves out of graves, a fact that William of Newburgh's stories support. The protovampires just tend to emerge from the grave. This hints that early proponents of the vampire myth might have been making up this element of vampire behavior to explain something they were seeing in the world around them.

Today, if a person in a family falls ill with a contagious and potentially lethal disease, doctors usually have the knowledge to identify it, prescribe treatment, and suggest quarantine measures if they are needed. In the days before modern medicine, when understanding of infectious disease transmission was rudimentary, people exposed to lethally contagious individuals had a good chance of following their friends to the grave. But they would not have done so immediately. Viruses and bacteria take time to spread through the body before having noticeable negative effects. This delay, which is known in the medical community as the incubation period, varies with the disease and can range from hours for some gut and respiratory infections to years for viruses like HIV. In most cases, though, incubation for diseases is a matter of days.

Imagine what people in those days saw after a loved one died from a highly lethal and contagious disease, like tuberculosis or a nasty strain of influenza. First, those who had lived with the diseased individual would fall ill upon the completion of the incubation period and run a high risk of dying. Then, those who had tended to these diseased individuals would also fall ill and transmit the disease before dying. One death would follow another in a dominolike progression. In a morbid sense, these patients were literally killing their friends and relatives, but from their deathbeds rather than from the hereafter. However, because of the incubation period, it wasn't clear to anyone how the disease was being passed along.

Driven into a panic by plagues of contagious diseases, people desperately sought an explanation. This search for answers even appears in William of Newburgh's story: The monster "filled every house with disease and death by its pestiferous breath." People were already somewhat aware of what was going on, but rather than pointing the finger at microscopic pathogens (which would have been impossible since microscopes were not even in use until the mid-1600s), they came up with the idea of the dead returning to kill off their friends and family. This led someone at some point to open up a grave and have a look. Shocked by the discovery of a bloody-mouthed corpse with a bloated belly, claws, and fangs, a connection was likely made between this horrific sight and the plague of death spreading throughout the community.

This seems logical, but it raises a question about timing. Why does the fear of vampires begin during the 1100s, when William of Newburgh was writing? Highly lethal and contagious diseases were hardly new things. In fact, Ian Barnes at Royal Holloway, University of London, published a study in the journal *Evolution* in 2011 revealing that infectious disease has played a key role in human evolution for centuries. Remarkably, this study found that humans who have been dwelling in places where population densities have been high for a long time carry genes that are particularly good at granting resistance to certain contagious diseases. This makes sense since places with higher population densities would have more humans available (and living in closer proximity) for diseases to infect and thus tend to be reservoirs where the infections could linger for long periods.* People who carried genes that coded for immune systems strong enough for them to survive this pathogenic onslaught proliferated while those who did not, died out. The study specifically notes that people from Anatolia in Turkey, where dense settlements have been around for nearly eight thousand years, carry a gene granting an innate resistance to tuberculosis, a disease that wreaked havoc in ancient cities. In contrast, people with almost no history of dense urban living, like the Saami from northern Scandinavia and the Malawians in Africa, do not show similar genetic resistance.

So it seems unreasonable to argue that people started digging up graves and inventing vampires as monsters only to explain the spread of contagious disease. If this were the whole story, vampires would be expected to have emerged as monsters much earlier. There had to have been other factors associated with the rise of the modern vampire, and clues to what these might have been can be found in one of the stranger vampire traits.

The sweet smell of garlic

According to some folktales, vampires are repelled by garlic, and for the most part this idea has remained tethered to the monsters for

*From a survival perspective, a bacterium or virus is in serious trouble if it finds itself in a small and isolated population. The disease will either kill off everyone and then die too when no hosts are left to infect or, under sunnier circumstances, everyone in the population will catch the disease, survive, and develop resistance to the disease so they never catch it again, a situation that also often destroys the disease. Pathogens depend upon having large numbers of people available to move through. This is why international travel is as much a boon to the diseases of the world as it is to economic development.

centuries. While modern enthusiasts of Gothic horror accept this trait as simply part of what vampires are, if you stop to think about it, being repelled by garlic is a rather bizarre quality to associate with a monster. The threat of sunlight makes the most sense. Evil things tend to be active in the dark and thus sunlight should naturally harm them. However, early vampire lore does not present sunlight as a threat. It is garlic that gets mentioned.

Garlic has a history of being used to protect the innocent from the forces of evil. The Egyptians believed that it could repel ghosts, and in Asia, garlic has long been smeared on the bodies of people to prevent them from being targeted by the spells of witches and wizards. Is there logic to this?

Some studies have shown that garlic fights infection, reduces blood pressure, and lowers cholesterol. For this reason, you could argue that any monster conjured up to explain inexplicable diseases, including vampires, came to be viewed as "fended off" by garlic because it was helping to boost immune system function. However, there is a problem. The scientific community is nowhere near any sort of consensus on the powers of garlic because its effects are, at best, weak. So one has to wonder: If modern researchers testing garlic's potential in controlled laboratory settings are having trouble determining if it really grants substantial benefits, were ancient people able to detect benefits at all? Or was there something else going on?

Foul odors created by corpses were often covered up by powerful smells like that produced by garlic, and there is some literature suggesting that, along with strong-smelling flowers, garlic was used at funerals where the corpse was getting a bit stinky. This may have been how it came to be connected with protecting people from the evil and the walking dead. Yet a most intriguing explanation for garlic being associated with vampires stems from the field of neurology.

During the 1600s, many Romanians believed that rubbing garlic around the outside of the house could keep the undead away, that holy water would burn them like boiling oil, and that throwing a vampire's sock into a river would cause the menace to enter the water searching for the sock and be destroyed.* Intriguingly, Juan Gómez-Alonso, a neurologist at the Hospital Xeral in Vigo, Spain, points out that these three things all have a connection to rabies.

In a report published in the journal *Neurology* in 1998, Gómez-Alonso explains that while the rabies virus can cause animals to become increasingly paralyzed as it spreads, in afflicted humans it can have a frightening effect on the mind leading to a condition known among medical practitioners as furious rabies. As the virus attacks their nervous system, patients become restless; some leave their beds and wander the surrounding area. They have trouble swallowing, frequently drool bloody saliva, become fiercely dehydrated and very thirsty. Worse, they often suffer from persistent feelings of terror and have a tendency to become angry and aggressive. Most important, furious rabies frequently attacks the section of the brain controlling how the body manages emergency respiratory activities like coughing and gasping.

Nerve cells in the lining of the nose, throat, larynx, and windpipe become extremely sensitive to noxious fumes and liquids. For this reason, patients with furious rabies suffer from spasms and extreme fear when they are forced to endure exposure to pungent odors (like that of garlic) or are presented with water (remember, they are desperately thirsty but cannot swallow). What do these spasms look like? When confronted, rabies patients tend to make hoarse gasping noises, clench their teeth, and retract their lips like animals.

Rabies has another connection to vampires based upon the way it is transmitted. Unlike, for example, influenza and tuberculosis, which are spread invisibly by particles in the air, rabies is primarily transmitted through bites. Most infections in people occur when a

---

*Romania is of course home to Transylvania and thus the epicenter of all things vampiric. At least, that is what the tourism industry would have you believe. Historically, though, there is nothing that connects Romania any more tightly to the origins of vampires than England, Belgium, or Hungary. It just happens to be the place that Bram Stoker chose as Dracula's homeland.

rabid animal breaks human flesh with its teeth and contaminates the wound with the infected saliva. The animals that most commonly spread rabies to humans in this way are dogs, wolves, and bats, all of which have a history in legends of being associated with vampires (bats are more recent than dogs and wolves, but all have been connected to the monsters for a while). Human-to-human transmission of rabies is all but unheard of today; however, historical accounts of people being bitten by rabid individuals do exist, and it seems likely that incidents of authorities or doctors being bitten while trying to subdue or capture rabid patients have taken place. In this case, the bite wound would heal as the rabies virus incubated inside the newly infected person's body. The individual who made the bite would die, but in time a new monster would be born.*

Rabies is spread not only through bites. It can also be spread through sexual activity. Furious rabies can cause hypersexuality and leave people with powerful feelings of sexual excitement. Men with the condition can develop erections that last for several days, and one individual is documented as having had sexual intercourse thirty times in a twenty-four-hour period before the disease claimed him.† With such powerful sexual stimulation at work and with patients so severely mentally compromised, it is hardly surprising the rabies literature reports violent rape attempts being common.

However, as tempting as it might seem to make a direct connection between rabies and vampires, rabies is very much a disease of the living and does not suggest that anything is returning from the grave. This does not disqualify rabies from being involved with the evolution of vampires. The virus probably did inspire the concept of vampirism spreading via bite and then merged with the perception that vampires were bloodsuckers. It is the undead element of vampires that rabies does not resolve, but, as mentioned earlier, there are

*Sounds almost like something out of a horror movie. Oh wait . . .

†In case you are curious, the sources that document this do not specify whether this was with one partner or many.

many contagious diseases, like tuberculosis and influenza, that could explain how people came to believe that the dead were returning from the grave to claim their loved ones.

In the end, the fears that ultimately led to the rise of vampires as they are known today may have come from people trying to make sense of two disease epidemics that took place roughly simultaneously. Tuberculosis was at epidemic levels in Europe throughout much of the 1700s just as a major rabies epidemic struck the wild dogs and wolves of the region. In one case, in 1739, a rabid wolf in France bit seventy people, and in another case, in 1764, forty people were bitten. To what extent these bites led to cases of rabid people biting one another is unknown, but if such a situation did develop at the same time as a town was suffering a tuberculosis epidemic, fears from each medical condition could have become intertwined. Even so, the idea of a curse turning man into a monster appeared long before vampires.

Bestial origins

Rabies, like influenza and tuberculosis, is well known to have been around since the dawn of humanity. Studies analyzing the evolutionary history of the virus show that it has existed in its present form for thousands of years. And before it came to be associated with vampires, the virus probably was part of the werewolf myth. The suggestion that vampires might have actually evolved from werewolves will certainly irk the fans of the *Underworld* and *Twilight* stories, where vampires and werewolves are mortal enemies, but a close look at the earliest literary descriptions of werewolves makes it tantalizing to consider a connection.

The *Satyricon,* believed to have been written by the Roman Petronius around AD 64, contains one of the earliest descriptions of a werewolf transformation. The tale is told by a man named Niceros traveling along a road with a soldier he has recently met. Along the way they stop in a graveyard at night. "We set out about cock-crow, the moon was shining as bright as midday, and came to where the tombstones are. My man stepped aside amongst them, but I sat down, singing, and commenced to count them up. When I looked around for my companion, he had stripped himself and piled his clothes by the side of the road. My heart was in my mouth, and I sat there while he pissed a ring around them and was suddenly turned into a wolf! Now don't think I'm joking, I wouldn't lie for any amount of money, but as I was saying, he commenced to howl after he was turned into a wolf, and ran away into the forest." Not long after, Niceros finds that all of the sheep at a nearby farm have been slaughtered by a wolf. Later, the soldier, now transformed again into a human, is discovered unconscious and under the care of a doctor.

So, a man strips naked, pees in a circle around his clothes under the light of the full moon, and transforms into a wolf. Was this a real wolf? Or was the man just snarling and howling as he lost some sort of mental control? The latter seems more likely in the event of a rabies infection, but the mention of a ferocious wolf attack could easily have been the work of a rabid wolf in the area. That the man is later found in bed under a doctor's care suggests he is actually ill. This hints that the werewolf as a monster may be a simplified version of the vampire before the fear of rabies became blended with the fear of other diseases.*

There is another side to all of this, though. In poor regions, where bodies were buried without caskets in shallow graves, it was not uncommon for wolves, which act as both predators and scavengers, to dig up the graves. They would eat human remains and, if caught during their feast, be thought by terrified witnesses to be

*Rabies is not the only condition that could have led the soldier to behave in this way. There are numerous drugs that can make people act like beasts and then fall ill. There is also a condition known as "excited delirium" that is starting to be recognized by medical communities. Sufferers strip out of their clothes, snarl and grunt like animals, lose control of their actions, become resistant to pain, and struggle fiercely when confronted. What causes the disorder is, to date, unknown but, like drug overdoses, it does not appear to be transmitted by a bite.

the exhumed person transformed into a creature of the night. This might, in fact, be why some early vampire stories describe the monsters as being able to take wolf form, and why, during the 1100s, William of Newburgh specifically mentions a pack of dogs following the monster as it spread death around town. Although, in some cases, and this will no doubt warm the hearts of the previously irked *Underworld* and *Twilight* fans, these wolf-scavenging activities also led to the creation of folktales suggesting that wolves were the sworn enemies of vampires and stayed near cemeteries to attack them as they tried to rise from the grave.*

The fears behind vampires and werewolves are very much the same. With both monsters there is the transformation of a relatively mundane human into a killer. On the face of it, this fear of a human becoming a predator is similar to the fears behind the Nemean lion, but it is taken a step further. In ancient Greece, lions were nocturnal hunters often not seen until it was too late. However, they were not common in towns, and people often felt safer near their homes. Werewolves and vampires made the monster human and, to a reasonable extent, allowed it to move among us disguised as a mortal. So it seems plausible that werewolf and vampire fears stemmed from more than just the threats presented by wild animals. They might also have come from fear of other humans.

Murder is as old as humanity itself. While Western audiences are most familiar with the biblical tale of Cain and Abel, this story is far from an isolated one. Tales of murder are central to the ancient myths of all societies; the fear of being murdered was very much a reality for many people in ancient communities. So were early werewolf stories a way of expressing this fear? In the tale of Lycaon, in Ovid's *Metamorphoses,* the connection seems obvious.

Lycaon, in an attempt to challenge the gods, presents Zeus with

*Yes, the antagonistic relationship between the werewolf Jacob and the vampire Edward in *Twilight* has a potentially scientific basis surrounding the scavenging behaviors of wolves in graveyards.

a platter of the chopped-up entrails of a person he has murdered. He doesn't tell Zeus this, though. Instead, he lies and declares it animal meat. Of course, with Zeus being a god, the ruse is detected. In a fury, the god slays Lycaon's sons with thunderbolts and curses Lycaon as he flees. Ovid writes, "Terror struck he took to flight, and on the silent plains is howling in his vain attempts to speak; he raves and rages and his greedy jaws, desiring their accustomed slaughter, turn against the sheep—still eager for their blood. His vesture separates in shaggy hair, his arms are changed to legs; and as a wolf he has the same grey locks, the same hard face, the same bright eyes, the same ferocious look."

As tightly linked as murderous behavior is to the werewolf in Ovid's story, there is a question of timing. Ovid was born in 43 BC and is thought to have died in AD 17 or 18. Murder, needless to say, existed before then. It is possible that, because of relatively low population densities, the unique mix of fears associated with rabies and violence did not merge until this time. However, it seems more likely that these fears and the horrible tales they inspired were just not written down and preserved until *Metamorphoses* and *Satyricon*. Certainly, the Mesopotamian *Epic of Gilgamesh,* written more than seven hundred years earlier than these two works, relates a transformation, initiated by the gods, of a man into a wolf. This might be early evidence of a fearful awareness that people in the community could, under certain circumstances, become as dangerous as predatory beasts.

Dawn of a new day

Regardless of their origins, vampires took a distinctly new form in Bram Stoker's *Dracula* in 1897. Passages from the book, like those found in Mina Murray's journal about her ill friend Lucy, show a connection between vampires and disease to still be present: "Lucy seems to be growing weaker, whilst her mother's hours are numbering to a close. I do not understand Lucy's fading away as she is doing. She eats well and sleeps well, and enjoys the fresh air, but all the time the roses in her cheeks are fading, and she gets weaker and more languid day by day. At night I hear her gasping as if for air."

Dracula, with his cruel, cunning, and elusive ways, took vampire fears far beyond those associated with disease. By linking Dracula to a coffin rather than a grave, Stoker made him remarkably mobile and thus capable of migrating from distant lands to places that were very familiar to his readers, like London. By being made fiercely intelligent and wealthy, he was very different from early vampires that were crawling out of graves. Dracula could buy whatever he wanted and manipulate those around him in the most subtle of ways. The vampire no longer had to scavenge like a ghoul; he could seduce the beautiful young women of the upper class. Instead of a mindless zombielike creature rising from the cemetery in a far-off land, the threat came in the form of a ruthless and brilliant murderer mingling with all classes of the city. It is a concept that must have been acutely terrifying to Victorians who were uncomfortably familiar with serial killers.

Yet above all else, Stoker made Dracula a charmer. He was eloquent, aristocratic, and exceptionally good at winning over women. This element of Stoker's monster may well have had a connection to the hypersexuality associated with rabies, may have stemmed from fears of rapists, or may have simply been an attempt to play upon societal fears of innocents becoming sexually corrupted in cities through manipulation. Regardless, the result was the construction of a monster that chilled readers to the bone.

Today, werewolves, vampires, and zombies still hold a deep fascination for society on the whole. Books like Stephen King's *'Salem's Lot* and movies like *Zombieland, Lost Boys,* and *The Wolfman* (among many others) tie into the ancient fears of deadly corruption being contagious. Although rabies is no longer as much of a threat to the Western world as it once was, contagious disease certainly is. With the SARS, swine flu, and avian flu threats that have struck society in recent years, fear of disease is very high. Anyone can be a carrier . . . the plumber, the cabdriver, the waitress at the coffeehouse, a spouse, a child. They mean no harm, but in this modern age of emerging diseases, everyone is a threat, bringing us very close to the concept of the innocent who has become infected with vampirism. This could be the reason for the continued appearance of vampires and vampire-like monsters that spread their curse as an infection.

But there is more. Diseases have appeared in recent decades that, because of their ability to make people lose control of their minds, are somewhat similar to rabies and the poisons used by zombie makers. Dementia and the hardening of the arteries in the brain are well known to cause mental malfunction, but among the most dreaded is Alzheimer's disease, where the mind is slowly attacked, memories are destroyed, and identity is stripped away. Worse, Alzheimer's causes people to forget not only their friends and family but also social rules. Indeed, it is common for sufferers of the disease to lose their inhibitions, make rude comments, and exhibit sexually inappropriate behaviors.

Admittedly, Alzheimer's and other disorders that lead to mental degradation are not caused by a bite or even airborne particles, but the fact that they are becoming ever more common makes the fear of losing control of oneself a very real threat. Fears associated with these diseases are likely leading to a certain level of modern vampire evolution.

In the novel *Harry Potter and the Prisoner of Azkaban*, the Dementors, which guard the wizard prison Azkaban, are ghostly creatures that cast a chill over all that surrounds them and attack by feeding on happy thoughts. They do not bite with fangs or drain blood; rather, they drain away their victims' happiness. Alfonso Cuarón's cinematic portrayal of them is distinctly vampiric in nature—they make sucking noises as they hover over their victims and siphon off happy memories.

While Dementors seem somewhat undead in nature, what is scary about them is not so much that they might have returned from the grave but that they can damage the mind to the point where a person is left with nothing but his or her worst experiences. They leave their victims as husks of their former selves, tainted individuals who will never again be truly human. This is not much different from what happens to the victims of vampires who lose their humanity as they rise up as undead themselves, hinting that although the Dementor appears to be an entirely new monster, to some degree it is the vampire adapted to play upon the fears of today's audiences.

Twilight years

Much of the discussion of how modern vampires terrify people leaves out a huge chunk of modern vampiric representation that is as popular as ever—the role of vampires as heroes. *Twilight, Buffy the Vampire Slayer, True Blood, Blade,* and *Interview with the Vampire,* to name just a few, all contain vampires that are partially or entirely good. Why establish a reversal from monster to hero? Nobody is putting Medusa, Chimera, or the Minotaur in the role of hero today. Why vampires? Part of the answer may be related to an ever increasing awareness of how easily corrupted we are that is leading us to be fascinated by heroes who must relentlessly fight their inner demons.

Unethical human behaviors are scientifically better understood today than ever before, and it is becoming increasingly obvious that there are not bad people and good people in society. There are just people who can, under specific conditions, be driven bad. As a zoological example, researchers can look at the behaviors of other species and predict what sorts of circumstances might make a typically helpful and devoted member of a couple give up on family responsibilities. Some marvelous work by Judith Morales and Alberto Velando at the University of Vigo in Spain and Roxana Torres of the National Autonomous University in Mexico looked at blue-footed booby males and worked out the precise conditions under which some males stayed to help look after the eggs and chicks that they fathered and others left the females to do all the chick rearing on their own.

The degree to which a male booby involves itself in family life depends upon two factors: how blue the feet of the female are and how large and colorful the eggs are. Females with very blue feet are deemed healthy because vibrant blue means they are eating well and not particularly stressed. Large and colorful eggs are also considered a positive sign of health. You'd think males would stick around with healthy females raising healthy chicks, but you would be wrong. Males give up on family participation under these conditions, presumably because they know they are leaving behind a family situation that can manage itself. However, males also give up when they discover that their mates have inadequate foot coloration and have laid poorly colored eggs, presumably because they figure the chicks are doomed and there is no way for them to make a difference. So when do males actually help? When things are bad, but not too bad, meaning the mother has poorly colored feet *or* the eggs are unhealthy-looking, but not both.* Something about this mix of stimuli triggers a damage control mechanism in the male's head that leads him to stick around and do what he can to look after the young. And it isn't crazy to consider that fatherly behavior in humans might be somewhat similar. The deadbeat dad phenomenon could very easily be a simple matter of some fathers being exposed to specific conditions that trigger a difficult-to-control evolutionary response.

Psychologists are also conducting studies to determine what sorts of conditions must be present for a normally law-abiding member of a community to engage in criminal activity. Rather remarkably, a 2008 study led by Kees Keizer at the University of Groningen and published in the journal *Science* found that people littered and trespassed far more often if they were placed in an environment where garbage and graffiti were present than if they were in a clean environment. Yet the effects generated by the filthy environment were not limited to these minor transgressions. During one part of the study, Keizer's team secretly monitored people who discovered an envelope

---

*How do booby researchers test this stuff? They kidnap females when the males aren't looking and use crayons to color their feet gray. Seriously, folks, this is cutting-edge science in action.

with cash visible inside it (there was a small window in the envelope) that the researchers had placed sticking out of either a tidy and clean mailbox or a mailbox covered in graffiti and surrounded by litter. The envelope was stolen only 13 percent of the time when it was in a clean mailbox, but that figure rose to 27 percent in the filthy conditions. And these findings are not alone. A 2010 study led by Chen-Bo Zhong at the University of Toronto revealed that people are much more likely to behave selfishly when it is dark and they are unlikely to be seen. Similarly, a 2012 study led by Shaul Shalvi of the University of Amsterdam found that when people are asked questions, they are far more likely to lie if they must answer quickly than if they are given ample time to think things over. Indeed, just as the booby equivalent of deadbeat dad behavior is triggered by simple stimuli, it is becoming increasingly obvious that people tend to behave badly when specific conditions are present. In short, we are becoming sensitized to the idea that we are often in no more control of our lives than animals. And if a filthy alley can lead a person who would not otherwise steal to engage in acts of theft, and the wrong foot and egg colors can lead booby males to abandon their families, what conditions need to be present in a marriage to lead a husband or wife to enter an extramarital affair or resort to domestic violence?

Is a rising awareness of just how easy it is for good people to go bad increasing our interest in heroes who are battling their own bestial natures in a very visible way? Vampires, it seems, display this beautifully. Certainly Edward Cullen in the *Twilight* saga fits this mold by choosing to feed only on the blood of animals and to befriend, rather than eat, Bella, the girl he finds so fascinating. Bill in *True Blood* behaves similarly in the first episodes of the series by spilling his own blood to save the life of Sookie. Angel, the love interest in the *Buffy the Vampire Slayer* television series, is in the same position, passionately kissing a woman whose protective crucifix burns his flesh. Even vampires who are sort of confusing to place as decidedly good or bad, like Louis in *Interview with the Vampire,* who refuses to take human life and chooses to feed on the blood of sewer rats for the first half of the story, endure this struggle. Moreover, many of these vampires experience pivotal moments where they are forced to duel with the monsters they truly are. To save Bella after she is attacked by another vampire, Edward must drink her blood and siphon away just enough of the evil vampire's essence to keep her from being corrupted but not drink so much that she dies. Similarly, Louis fights his desire to feed on the servants of his plantation, trying—and failing—to protect human lives from his own nature.

An explanation of why vampires are being viewed more positively over time could also be similar to the reason why snakes turn up in mythology as both monsters and godlike creatures of creation and fertility. As discussed with dragons, in spite of their long history of killing people, snakes have sometimes been viewed as something akin to divine because they seemingly survive off of no food and appear to stay perpetually young by shedding their skins. It is possible that the element of immortality that has woven its way into the modern vampire myth is causing perceptions of vampires to function in a similar way. They evolved from a long history of fear of contagious disease but have now attained a form in the human imagination that possesses a trait many today covet. Just as snakes historically have been both feared as monsters and worshipped as symbols of fertility, the vampires' role in society is changing as fear of them turns to enthrallment with their status as undying creatures. Could a key reason why vampires are now viewed as sexually attractive heroes be that they are presented as lovers who know no pain and who will be eternally young? It seems plausible, but there might be another element at work.

Vampires, as they have been known for so long, might now be entering their twilight years as the result of increasing scientific awareness. People knew next to nothing about communicable diseases when vampires first emerged as monsters, but this is no longer the case. When swine flu broke out, epidemiologists were quick to spot where the disease was appearing, how it was spreading, and, most important, how to contain it. Avian flu is an even better example. Researchers around the world are aware that the virus can, under rare circumstances, leap from birds to people and that some strains of avian flu inflict high mortality. This is why the disease is getting so much attention and why monitoring stations have been set up worldwide to keep track of how the disease is behaving (and evolving) in other animals.* So fear of contagious disease is as present as ever, just as it was with the rise of vampires, but there is now a major difference: The vast majority of the population understands where the true monster resides.

Thrillers like *The Andromeda Strain, Outbreak, The Hot Zone,* and *Contagion* all give audiences a white-knuckle ride through the panic that could realistically strike if a horrific disease were to emerge. There is no question that the monster in these films is the disease, but this fact raises an intriguing question. Are these disease monsters merely a new form that the fears responsible for the rise of vampires are starting to take as scientific understanding spreads?

In *Interview with the Vampire,* the ancient vampire Armand comments, "The world changes, we do not. Therein lies the irony that finally kills us." The point is that a vampire's inability to cope with the ever-changing human world eventually leads to the vampire's destruction. However, when taken in the context of vampires and their life span in society as monsters, the statement could not be more true. The world is changing, and vampires as they have been known for so long may soon no longer find a place in it. Their only choice seems to be to evolve into heroes that feed on animal blood, like the Cullens in *Twilight,* or on artificial blood, like Bill in *True Blood.* To do otherwise may be to face extinction.

---

*One of the coolest monitoring systems is in zoos in the United States. Because zoos have a wide variety of animals that are very closely watched by veterinarians, and because these animals have good access to wild birds (think pigeons and ducks), if a virus evolves that is capable of jumping from birds to other animals, it is likely to be spotted first in a zoo, where experts are now watching.

8

The Created—The Golem, Frankenstein, HAL 9000, Terminator

"I'm sorry, Dave, I'm afraid I can't do that."

—HAL 9000, *2001: A Space Odyssey*

The idea of life being created by the hand of humanity is hardly new. The mythological Greek inventor Daedalus was famous for constructing wings that allowed him to fly out of the labyrinth he designed for the imprisonment of the Minotaur, but these wings were not his only invention. Classical writers described other wonders he built, including statues with lifelike qualities. The only evidence of these inventions having existed comes in the form of notes made by Plato in *Meno*: "The images of Daedalus . . . if they are not fastened up they play truant and run away; but, if fastened, they stay where they are." Daedalus, it seems, had found a way to bring his sculptures to life, and if they were not chained to a wall, they ran off.

Yet Daedalus's mechanical creations did not threaten anyone.

They were curious and odd but never found harming humans, and it is this point that defines them as wonders rather than monsters.* While many creatures "created" by the hands of gods turned out to function as monsters—Medusa, Scylla, and Chimera, for instance—there are no classical texts suggesting people ever created such things.

By the 1500s, this was no longer the case. Jewish mythology tells of a rabbi who fashioned a human-shaped structure out of clay and then made prayers that allowed it, through divine intervention, to function as a living creature. The story reportedly existed as oral tradition for centuries, but in 1909 a manuscript was found that was claimed to have been written by the Golem-creating rabbi's son-in-law.† This document detailed how the creature, which came to be known as the Golem of Prague, was created: "I continued circling the glowing figure in the mud, and soon lost awareness of the number of times I had encircled it. The outline of the Golem was glowing more and more intensely and the odor of scorched earth reached my nostrils . . . no sooner had we completed our slow, concentrated recitation of the verse in Genesis, when the Golem opened his eyes."

The role of the Golem is well outlined in notes. The rabbi ordered it to protect the Jews of Prague from the constant persecution they were enduring from Christians, and it fills this role perfectly. The tales that are told in this work (including one harrowing story where the Golem jumps aboard a rushing wagon and wrestles with a man who is about to frame a Jew for murder) represent the creature as nearly invulnerable, incredibly strong, and always fighting to bring criminals to justice.

The Golem is undeniably alien in form, but classifying it as a monster is difficult because its behaviors are not outwardly harmful toward

---

*This is not to say that stories portraying them as dangerous will not one day be discovered; they have simply not been found yet.

†There is a lot of debate over how old the Golem story really is and whether the 1909 text that was "found" actually had been written by the rabbi's son-in-law. Some argue that the "finder" of this work, Yudl Rosenberg, wrote the text himself to fantasize about fighting the increasing violence that Jews were facing during the early 1900s.

humanity. It does attack people, but we can all agree these people are villains. Does an aberration that regularly causes harm to evil humans count as a monster? Or does it count as something else? Technically, the mutants of the *X-Men* comic books and movies are also aberrations who frequently harm villainous humans. Are they monsters? In the comic universe within which they exist, they are not. Really, the Golem is more of a historic clay superhero than a monster.

From a believability standpoint, it is hard to see how anyone could think that a pile of wet clay can become animated by being shaped into human form and exposed to prayer. There are places in the world where clay can take on lifelike qualities, bubbling thermal pools being one of them, but the concept seems too far-fetched. The stories of the Golem do not have the creature sneaking about in places where people could have mistaken some natural event for a creature of this sort. The Golem's activities always took place in towns or inside homes, and were easily seen during the day. So it could not have been a case of mistaken identity. What is possible is that there was a vigilante at large in Prague at the time this tale was recorded, something akin to a Jewish *Dark Knight,* and that the Golem mythology was invented as a form of cover story much as the masked Batman was invented as a cover for Bruce Wayne. For all we know, the rabbi who created the Golem, or his son-in-law who supposedly recorded the tale, could well have been the vigilante himself.

The story of the Golem of Prague ultimately concludes with anti-Semitic behavior coming to an end. The rabbi, seeing that all is finally well, declares, "I feel the golem must be destroyed. His mission is accomplished. If we keep him longer than necessary, we may fall into the danger of misusing sacred property." And with this sentiment, the creature is done away with in prayer.

From synagogue to laboratory

If *The Golem of Prague* is something of a cautionary tale about creating artificial life only when the need is great and God consents, the

1818 publication of Mary Shelley's *Frankenstein* was the cautionary nightmare. While a lump of human-shaped clay may have achieved only certain levels of horror, a creature built from scavenged body parts was another matter entirely.

Shelley's writing is infused with dread and awe: "It was on a dreary night of November that I beheld the accomplishment of my toils. With an anxiety that almost amounted to agony, I collected the instruments of life around me, that I might infuse a spark of being into the lifeless thing that lay at my feet. It was already one in the morning; the rain pattered dismally against the panes, and my candle was nearly burnt out, when, by the glimmer of the half-extinguished light, I saw the dull yellow eye of the creature open; it breathed hard, and a convulsive motion agitated its limbs."

This contrasts rather starkly with more recent presentations of *Frankenstein* that portray Victor Frankenstein standing over a lab table hysterically shouting "It's alive!" as lightning bolts strike and send electricity surging through the body of his creation. The original tale contains nothing of the sort. There is no lightning, no hysteria, and the story is all the more frightening for it.

The scene is dark, with the only candle nearly out. The rain is spattering against the windows and it is very late at night. But while the setting itself establishes fear, it is the eye that immediately establishes the creature as an aberration. Human eyes come in many colors, brown, blue, green, nearly black, but yellow is not an option. By specifically choosing a color impossible for humans to have, Shelley places the creature into the category of nonhuman. And she goes much further: "His yellow skin scarcely covered the work of muscles and arteries beneath; his hair was of a lustrous black, and flowing; his teeth of a pearly whiteness; but these luxuriances only formed a more horrid contrast with his watery eyes, that seemed almost of the same colour as the dun-white sockets in which they were set, his shrivelled complexion and straight black lips."

Like Chimera, a lot of the fear generated here stems from asymmetricality. Perfection and rot side by side is dire stuff, but grotesque descriptions could take the horror of Frankenstein's creature only so far. Telling somebody about a dreadful sight is nowhere near as frightening as suggesting that such a horror is real and potentially nearby. Because of where medical science was headed during Shelley's day, she was able to make such suggestions very easily by tapping into recent developments in blood transfusion research.

Blood transfusions are a common part of modern medical treatment. If a patient loses a lot of blood, through either a traumatic accident or surgery, it is standard practice to replace the lost blood with blood that has been donated by other individuals. This sounds relatively simple, and today it mostly is.

Doctors know that the patient's blood type as well as the donor's must be identified and matched before transfusion. Failure to do this is extremely dangerous. The body has extensive defensive systems that are very good at destroying foreign materials, including new blood, even if it is vital to the body's survival.

If infused blood does not match the patient's blood, the immune system aggressively attacks under the mistaken assumption that it is a foreign invader.* These reactions are devastating. They frequently lead to organ failure and can easily cause death. For this reason, during the early 1800s, when blood types were not understood, giving people blood from donors proved very risky. Even so, attempts were being made, and researchers in Britain were working out that blood did not need to be transferred directly from one body to another. During the 1820s, trial and error with women who were in desperate need of blood following childbirth revealed that blood could be placed into a syringe for a short time before being injected.

---

*Fascinating work conducted by the biomedical engineer Maryam Tabrizian at McGill University in Montreal and recently published in the journal *Biomacromolecules* shows that we are now able to effectively make red blood cells invisible. By coating the cells in thin polymer layers that still allow them to do their duties, Tabrizian and her team showed that it is possible to grant red blood cells of one blood type the ability to function inside a body that is used to red blood cells of another type. While currently being tested only in mice, initial results suggest that the technology works and that a day may soon come when worrying about having the right blood type for a patient becomes a thing of the past.

Thus, the crucial ingredients existed at this time for a belief that blood could be collected from a human, stored in a container, and then eventually used to sustain another person's life, even after the donor's death. Mind you, medical science was not far enough along to practically conduct this procedure in 1818, when *Frankenstein* was published, but the ideas were there. And if blood could be stored after death, why not body parts?

Today hearts, lungs, kidneys, and even, according to one recent study, tracheas can be moved from one person to another. During Shelley's day, this was not the case; if people received transplants, they died. But tests with animals were being conducted and a lot of information was accumulated. So it seems plausible that readers of *Frankenstein,* during its first decade of publication, believed that, just like blood, organs could be collected and reused after death. It is this connection to reality that must have made Shelley's monster not just a black-lipped, yellow-eyed horror but a science-made terror that could really be created at a research institute.

Sexual awakening

Frankenstein was only the beginning. Since the publication of Shelley's novel, creatures crafted by the hands of humanity have continued to find their way into monster literature. The beast folk in H. G. Wells's *The Island of Doctor Moreau* are very similar to Dr. Frankenstein's monster in that they are built by a scientist. Even though Dr. Moreau does not breathe life into them as clearly as Dr. Frankenstein does (they are living animals at the start and living animals at the end of the scientific process), he is effectively responsible for their creation. Dren, the monster featured in Vincenzo Natali's 2009 film *Splice,* is another example. Using modern science to make the story believable, Dren is created from animal DNA that is spliced together with human DNA rather than stitched together from a combination of animal and human body parts (a method that would lead to a severe immune system reaction that would make foreign blood rejection look tame). Dren is effectively a genetically engineered creature with an alluring feminine face, legs like those of a quadrupedal beast, and a tail with a stinger on its end. This combination places her in the unusual situation of being simultaneously frightening and erotic. It also makes her something of a modern Chimera. The utterly lethal monster Sil in Roger Donaldson's 1995 film *Species* is similarly created through genetic engineering, with scientists using genetic information sent to them from space to craft a new organism. Again like Dren, Sil is simultaneously sexually alluring and menacing. This is a far cry from Dr. Frankenstein's monster and Dr. Moreau's original beast folk, which were all entirely hideous, but it is not different from the modern adaptation of Dr. Moreau. The 1996 film version of Wells's tale, which deviates substantially from the original text, includes a female member of the beast folk named Aissa, who is both attractive and fanged. So why is sex appeal being increasingly merged with dangerous creations?

Part of the answer presumably has to do with money. Create a frightening monster, and people come to see it. Create a frightening monster with boobs, and even more people come to see it.* But there is more. Stitched-together corpses and animated clay structures are not the sorts of human creations being sexualized. The creations that are, tend to be formed by mixing human DNA with the DNA of some other species. With Dren and Aissa, the other species is a real-world animal; with Sil, the other species is an alien. In all three cases, the hybrid creature formed from the genetic blend is strong willed, sexually attractive, and more powerful than any normal humans.

It is going to sound absurd, but to come to grips with why sex is so strongly emphasized in Dren, Aissa, and Sil, it is worth considering the humble male peacock. Here is a bird that needs to grow feathers of ridiculous size and brilliant color to attract a female. The growth of these feathers demands that the male consume a lot of nutrients. Moreover, the feathers' mere existence presents the male with a con-

*By people, I mean men.

stant problem—they dramatically reduce mobility and make the bird more vulnerable to predators. Yet evolutionary logic dictates that any males that can survive such risks and still have huge and attractive feathers must be the best of the best and worthy of mating with. And the pressure is not only on male animals. Females of many species must also take care of themselves; just consider the blue-footed booby females that must keep their feet bright and blue or risk losing their partners.

People like to think of themselves as more highly evolved than peacocks and boobies but, honestly, the differences have not traditionally been viewed by psychologists as all that vast. Decades of studies reveal that humans are attracted to mates who present positive characteristics. Specifically, they show women to be keenly attracted to men who have money and power, while men are attracted to women who project health and strength, like good skin, symmetrical faces, and a favorable waist-to-hip ratio.* This all makes perfect sense. From an evolutionary perspective, women benefit greatly from breeding with men who have the resources to take care of them, while men benefit from breeding with women who are healthy enough to survive childbirth and pass along good genes to children.

As for where this leaves Dren, Aissa, and Sil, they are female creatures infused with genes that make them stronger and healthier than any other humans. They are symbols of the ultimate in physical attraction, but they come with warnings. Like Dr. Frankenstein's monster, the old warning arises not to meddle in matters of creating life, but there is also a new warning: Beware deeply seated evolutionary instincts. These new sexual females and genetically engineered monsters hint of the dangers associated with falling prey to behaviors driven by attraction based upon millennia of natural selection and not upon logical thought. Their presence in films suggests that something many people fear in these days of increasing

*Very important always to have a measuring tape and calculator on hand during first dates for such things.

psychological awareness is being lured by physical attraction into a relationship with someone who is predatory. It is a fear of the femme fatale.

Yet as the science of psychology moves along, monsters like Dren, Aissa, and Sil may ultimately find themselves going the way of the dodo. The psychologists Eli Finkel at Northwestern University and Paul Eastwick at Texas A&M University are experts in the science of romance. As they examined what it is that attracts males to females, and vice versa, they found flaws in many of the older research reports.

The volumes of historic psychology studies that helped form the widely held belief that men want gorgeous women and women want wealthy and powerful men were almost entirely based upon information gathered from people who were seeking dates or about to go on dates. This was not because psychologists were lazy; studying human dating behaviors is difficult, and the best way to do it is to ask people what they were seeking in a partner. It wasn't practical to follow people around as they went on dates and monitor what they said and did to one another. This changed with the invention of speed dating.

During speed-dating activities, one gender (usually female) is seated at tables around a room while the other gender (usually male) rotates, spending a few minutes with each woman before moving along. Afterward, participants select whom they want to see again, and if the person they selected chooses them too, the speed-dating system reveals contact details to the interested individuals and they can set up an actual date.

Eastwick and Finkel have effectively hijacked speed dating in the name of science. By offering speed-dating events with video cameras placed all over the room, recording systems on every table, and participants given a discount for consenting to have their activities captured, the duo have learned much. Specifically, they have found that what women and men say they want is often not what they really are attracted to when presented with actual people.

"It is like meeting someone who says she doesn't like eggs but who then goes on to have cake made with egg but not complain at all. A woman may have a preference, but that preference may be entirely ignored when a complex person is presented in front of her," explains Finkel. "The same is true of men. Our findings are showing that in live interactions, men care about finding someone who is personable and has good earning potential just as much as women do, they just don't express these preferences when asked," adds Eastwick.

The work is relatively new to the field and it will take time for it to become widely known, but as it does, it will likely erode the fear that we cannot resist the allure of an enticing but dangerous lover, and this, in turn, will likely drive the sexually terrifying aspects of creations like Sil and Dren to extinction.

Born to be bad?

Sexual aspects of monsters created by humanity aside, as terrible as Dr. Frankenstein's creation might have been to look at, it is not just its looks that make it frightening. The creature commits several murders, becoming the terror that everyone initially feared. But it is worth asking whether it was actually "born" a monster.

Many of the early monsters, like Chimera and the Minotaur, were born both ugly and evil. This is not the case with Dr. Frankenstein's creation. The creature is not initially a threat. It is most certainly a hideous aberration, but during its first months of life, when it has not yet attacked anyone, it is difficult to classify as a monster since this, by definition, requires it to cause harm to people, and it doesn't.

During its early days, the creature finds a hiding spot inside an old barn and observes a family on the outskirts of a small town. Without intruding, it studies them and develops a keen yearning to be accepted. The monster starts helping the family by chopping wood and shoveling snow at night, giving them the impression they have a guardian angel looking after them. In time, the creature approaches the family house when everyone but the blind grandfather is away. It speaks to the old man, telling him it has come to visit friends who have never seen it and that it is fearful of their reaction. The old man consoles the creature, but to no avail. When the younger members of the family return, they react badly, attacking the creature or fainting in fear.

It is from this experience that the creature decides that the only way for it to enjoy companionship is if it kidnaps a young child, who does not yet know fear, and teaches the child to accept its wretched form. By chance, the child it kidnaps is Dr. Frankenstein's youngest brother. As the child screams that the Frankenstein family will avenge the kidnapping, the creature changes its plans and strangles the child as an act of revenge toward its creator. Not satisfied, it then frames a young girl in Dr. Frankenstein's hometown for the murder. It is here, and during the series of killings that follow, where the creature evolves from being merely hideous to being a monster.

In many ways, the situation for Dren in *Splice* is similar. After being created for research and carted off to a barn where she is hidden from public sight, she slowly begins to grow bored and despondent. While she might not be entirely composed from human DNA, Dren is inquisitive and intelligent. Yet she is locked away like a prisoner and treated by her creators as more of a pet than a person. Over time, her frustration grows. She first lashes out at a cat, piercing its body with the stinger in her tail. Soon after, she pins down her creator and steals the key she needs to escape. The conflict quickly grows out of control and Dren becomes a monster.*

Such discussion of Frankenstein's creature's and Dren's psycho-

---

*An intriguing aspect of this film is that Dren undergoes a sex change. While female for much of the story, she is most violent after transforming into a male and ceases to be a character we sympathize with. Pages could be written on the gender psychology being invoked, but I'm not going to get into that. I'll just point out that the biology here is all wrong. Although there are many animals that change gender during their lives, they do not usually switch from female to male; they switch from male to female. This all has to do with logistics. Being a male is cheap and easy. You just release sperm and let the female do the heavy lifting. Thus, if you are going to be a male and a female during one life, you want to be male first (when you are small and weak) and female second (when you are bigger, tougher, and have more resources at your disposal to produce offspring). But the *Splice* story wouldn't have worked if the gender swap had been done the other way around, and I guess Natali figured that most audiences weren't geeky enough to catch him getting the biology wrong.

logical journeys from innocents to killers might seem out of place when it comes to analyzing the role of science in the formation of their status as actual monsters, but it is not. To get at the fears underlying these creations, it is crucial to realize that the biological work responsible for making their horrific forms believable and terrible to behold plays only a partial role in their transformation into evil beings. The science wielded by Dr. Frankenstein and the researchers in *Splice* creates only hideous creatures. The infusion of evil into them depends upon their interactions with humans. In *Frankenstein,* the evil arises after Dr. Frankenstein's horrified withdrawal from his own creation, the violent response of the family in the countryside, and the heated words shouted by the young brother whom the monster kidnaps. With Dren, the birth of evil stems from her imprisonment and ultimate treatment as nothing more than a specimen.

Thus, there appear to be two fears crucial to the formation of these monster stories. There is the fear of what horrific things science is capable of creating, and then there is the more subtle fear of society's inability to recognize a creature's needs and react appropriately such that the creature does not become so wounded that it turns against humanity. There is also the fact that both Frankenstein's monster and Dren are physically very powerful. We all know that power corrupts, but in recent years researchers have discovered that giving a person a combination of high power and low social status creates a particularly horrific psychological effect.

In a study conducted by Nathanael Fast at the University of Southern California and published in 2011 in the *Journal of Experimental Social Psychology,* 213 participants were randomly assigned one of four situations that manipulated their status and power. All participants were informed that they were taking part in a study on virtual offices and would be interacting with, but not actually meeting, a fellow student who worked in the same fictional office. These people were later assigned either to the high-status role of "idea creator" and asked to generate important ideas, or to the low-status role of "worker" and tasked with menial jobs like checking for typos.

To manipulate power, participants were told there would be a draw for a $50 prize at the end of the study, and that, regardless of their role, each participant would be able to dictate which activities his partner must engage in to qualify for the draw. Participants who were given a sense of power were told that one part of their job required them to determine which tasks their partner would have to complete to qualify. They were further informed that their partner would have no such control over them. In contrast, low-power participants were advised that while they had the ability to determine the tasks their partner had to engage in, their partner could remove their name from the draw if he or she wanted to.

Participants were asked to select one or more tasks for their partner to perform from a list provided by the researchers. Some of these tasks were rated by a separate pool of participants as deeply demeaning, such as requiring participants to "say 'I am filthy' five times" or "bark like a dog three times," while others were deemed neutral, like "tell the experimenter a funny joke" or "clap your hands fifty times." Fast found that participants with high status and high power, low status and low power, and high status and low power all chose few, if any, demeaning activities for their partners to perform. In contrast, participants who were low in status but high in power were much more likely to choose demeaning tasks for their partners.

To a certain extent, these results provide a psychological explanation for the behavior of the prison guards at Abu Ghraib in Iraq. They were locked, loaded, and very high in power, but they were prison guards; they knew they were viewed by society as low in social status. Similarly, Fast's findings make the evil transformations seen in the socially excluded but physically powerful Frankenstein's monster and Dren all the more believable. We fear this type of transformation because it actually happens in humans all too often.

Of silicon and metal

It is shortsighted, however, to focus only on monsters spawned from biology. While Dr. Frankenstein's monster was a product of biological science, there are many recent monsters bearing a striking resemblance to these creations that are not flesh and blood.

Like Dr. Frankenstein's monster, the computer HAL 9000 in Arthur C. Clarke's *2001: A Space Odyssey* and Stanley Kubrick's film version of the story is created by humans. Represented by a single red eye in the film, HAL is found throughout the spaceship that it is meant to help run. However, during the mission, something goes dreadfully wrong. An electronic malfunction leads HAL to make a mistake and declare equipment to be in need of repair when it is operating normally. The astronauts grow concerned about HAL's error and consider shutting down the computer. They discuss this in private inside a small space pod that they believe HAL cannot eavesdrop on, but HAL, suspicious of the astronauts' behavior, reads their lips through the pod window and works out what they are planning. This leads HAL to start killing off the crew. The computer is able to rationally explain the reason for its murderous activities since it views itself as critical to completing the space mission, but rational or not, the way HAL snuffs out the lives of the humans on board is undeniably creepy.

In Andy and Lana Wachowski's 1999 film *The Matrix,* the sagelike character Morpheus comments, "We marveled at our own magnificence as we gave birth to AI [artificial intelligence]," as he explains to the protagonist Neo that this marvelous technology turned on humanity and effectively declared a war that it mostly won. In James Cameron's 1984 film *Terminator,* a similar plot unfolds, with intelligent machines invented by people rising up against their creators. Even television has carried this story, with the successful *Battlestar Galactica* series always opening with the bold lines "The Cylons were created by man. They rebelled . . . ," providing an explanation for why humans are constantly being chased by the robotic Cylons around the galaxy.

The machines in *The Terminator* and Cameron's 1991 sequel, *Terminator 2, Judgment Day,* have the same reasons for attacking humanity that HAL does. They become self-aware, humans attempt to shut them down, and the machines retaliate in self-defense. While Skynet, the artificial intelligence that controls the Terminator, and HAL are both bent on killing off humans, is it right to classify a species fighting for its survival as a monster? Are Skynet and HAL any more evil than a bear that tears the arms off of a hunter who just took a shot at its cub? To a certain extent, the answer is yes, because a bear mauling a hunter does not go off and start mauling every human it meets.* With Skynet and HAL, a paranoid logic develops in these systems that all humans, even those who are harmless, must be killed, and this is where the evil begins to seep in.

But why do monsters venture into the world of computers in the first place? Unlike a decomposing corpse or a deformed lab animal, computers are not inherently grotesque. HAL's lightbulb, on its own, is just a light. There is nothing inherently frightening about it. With *The Terminator, The Matrix,* and *Battlestar Galactica,* this changes somewhat as computers are given more physical form. The Terminator is an eerie-looking skeletal robot covered in human skin, the Cylons are large and powerful with weapons on their arms (or in some cases programmable humanlike machines), and the lethal programs that function as guardians of the computer system in *The Matrix* appear as spooky and dispassionate government agents.† But it does not seem that it is the physicality of computers that leads creative minds to transform them into monsters. It is what computers are capable of that drives this process.

A team led by Louis-Philippe Morency at the University of Southern California is showing that, when properly programmed and

---

*Contrary to popular belief, the bears that are actually the most dangerous are those that lose their fear of people by being fed by them. These animals come to see humans as a free meal ticket and become aggressive when they are not given the food they want and expect. Bears that get shot at do everything they can to stay away from people. Sure, we still read about tourists being ripped apart after trying to nuzzle a bear cub or photograph a grizzly at close range, but seriously, are these cases of the bear being dangerous or the human being stupid?

†Fear of the federal government. Totally irrational, right?

hooked up to video cameras, computers are becoming adept at reading human body language. More specifically, Morency and his colleagues have taught computers how to read the all-important human nod.

This might sound insignificant, but nods made at the right time in a conversation can mean "I understand," while nods made at the wrong time can indicate either a lack of understanding or a lack of interest. Teaching robots and computer avatars to identify these different sorts of nods and to properly nod back has been a nightmare because a definition of exactly when nods of different sorts are supposed to happen has not existed.

Psychologists have tried for ages to figure out the many subtle elements produced by a speaker that lead a listener to nod, and the results have been poor. To solve this problem, Morency turned to computers. By recording movements and sounds during human interactions, he has generated lists of conversation cues—like pauses, gaze shifts, and the word "and"—that lead people to nod. He has also collected facial details that indicate what sorts of nods are being made. This information is now being fed into computer programs and used to teach robots when to nod during conversation and what human nods really mean.

At the most basic level, Morency's work, and similar face analysis technology, could ultimately prove rather valuable for authorities keen to identify expressions associated with malice and deceit. But there is much more. As computers are required to interact with humans more often, their ability to understand everything that is communicated, rather than just typed or spoken, is going to vastly improve, opening up communication pathways so computers can start playing a larger role in social interactions. Imagine educational software that can detect the glazed look of someone who is totally lost during a lesson or a spaceship computer that suspects two astronauts are lying and reads their lips to learn what they are really thinking. This sort of work is a big step forward. And just in case you thought that all of the *Terminator* and *Matrix* films might have left people wary enough of artificial intelligence to keep such systems out of war machines, be assured that you are wrong.

A team led by the computer scientist Yale Song at the Massachusetts Institute of Technology is teaching military drones to understand how to read the body language of deck officers on aircraft carriers and follow their commands. The ultimate goal of the work is to have the drones read the silent signs and signals just as well as human pilots do. At the moment, the drones understand what they are being told only 75 percent of the time, but they are going to get a lot better as the work progresses.

Along similar lines, Andrew Gordon at the University of Southern California has designed computers that are adept at reading blogs and constructing meanings from what they find. For example, after scanning millions of personal stories online and correlating these with incidents taking place in the real world, his computers were able to work out that rainy weather was related to increased car accidents and that guns were associated with hospitalization. Connect these sorts of developments to computers that are capable of crushing the brightest humans at chess and beating geniuses on *Jeopardy!* and something emerges in the imagination that certainly gives one pause.

The day when a computer is capable of truly studying the world around it, learning from what it finds, engaging in flawless social interactions, and acting independently is not that far off. Just as Shelley's early readers were frightened by how far transfusion and transplant technologies could be taken, so too are modern readers frightened by what sort of form artificially intelligent computers will take. Really, the idea of self-aware war machines getting tired of being treated as servants or simply malfunctioning and going on a killing spree is not hard to imagine. And it is from this fear that robotic monsters arise.

Yet mixed with this fear is a lot of hope. In *Terminator 2,* a reprogrammed robot, physically identical to the robot monster in the first film, is sent back in time to protect the future leader of the resistance movement against the machines from assassination when he is just a boy. The boy alters the Terminator's programming, giving it the ability to learn. He orders it to not kill humans, teaches it to express emotion, and encourages it to question human behavior, leading to an unexpectedly tender moment where the machine asks him why humans cry.

In the end, the two build a bond, and the robot, which was a vile killing machine in the first film, concludes the second film by displaying an understanding of the value of human life and willingly sacrificing itself to save humanity.

Isaac Asimov, the creator of some of the most profound robotic literature ever written, presents a similar tale in the first story of *I, Robot,* "Robbie." The tale explores the social interactions that develop between a human child and a robotic nursemaid named Robbie. After two years of happy bonding, the pair are separated by the child's mother because she decides it is socially inappropriate for robots to become so closely attached to humans. This drives the child into a state of depression that leads her toward a desperate search for her lost companion. Toward the end of this search, she finds Robbie installed in a factory. As she rushes up to meet the robot, she fails to notice an oncoming vehicle. Robbie quickly saves her, proving to the mother she was wrong in believing robots to be cold and soulless.

"Robbie" raises a vital point that deserves some reflection. In this story, it is the mother, not the constructed creature, who is the antagonist. She is not a monster, but she is definitely the force the leading characters must struggle against. In *Splice, Frankenstein, Battlestar Galactica, The Terminator,* and so many other stories of human creations becoming monsters, humans might not be the antagonists, but they are definitely responsible for the monsters that come to haunt them. But when does a sheer lack of responsibility shift a character from simply being incompetent to being a villain? Can such monster-constructing villains become monsters themselves? It is with these questions in mind that it is worth taking a look at *Jurassic Park.*

9

Terror Resurrected—Dinosaurs

"The lack of humility before nature that's being displayed here, uh . . . staggers me."

—Dr. Ian Malcolm, *Jurassic Park*

While lions and bears were truly terrible threats to early humans, fossils show all too clearly that these predators were nothing compared to the animals that came before them. Yes, the media have relentlessly discussed dinosaurs in recent decades, but there is still a real lack of appreciation in society for how big and dangerous they actually were. The long-necked species *Diplodocus* is a good example. It topped out at around a 30 feet (10 meters) long and 20 feet (6 meters) tall, larger than many buildings. Carnivores were smaller. *Tyrannosaurus rex,* by no means the largest of the meat-eating dinosaurs, was only a bit taller and 50 percent longer than the largest giraffes. Of course, a statement like that is preposterous. A bit taller and 50 percent longer than a giraffe is still huge. Giraffes are big animals. Being right next to them is nerve-racking. One false step and they can easily break a human foot or, worse, snap a spine.

Based on size alone, the average *Tyrannosaurus rex* was much more of a threat than the long-necked plant-eater known so well today. Yet add to it the fact that its mouth was loaded with pointed teeth the size of bananas, and the results are mind-boggling.

It is difficult to know for certain just how strong the bite of a *Tyrannosaurus* was, but comparisons with modern animals can give scientists a rough idea. Numerous bear and great cat species have jaws strong enough to crush a human head like a grape. With this in mind, *T. rex* would have been able to deal out a whole lot more damage had humans been around when it was alive.*

Aside from being fun for paleontologists, trying to work out how an ancient predator like *Tyrannosaurus rex* functioned is a great exercise for realizing just how small modern predators actually are. Perhaps more important, it is also an opportunity to recognize why dinosaurs have featured so prominently as monsters in stories since their identification by the scientist Sir Richard Owen in 1841. Even in the Victorian era, when Sir Richard was doing his work, he must have felt that they were frightening animals, because he chose to formally call them "*deinos*," meaning "fearfully great," and "*sauros*," meaning "lizard."

Dinosaurs have not changed since their discovery more than a century ago. Their bones are still just as big and their teeth just as sharp. Yet they have risen and fallen through the decades as monsters. There have been periods of intense interest along with some lulls. Why?

A lot of this likely has to do with believability. The waxing and waning of dinosaur popularity seem to follow trends that question whether they might "still be out there" or potentially be "resurrected."†

Early books and films that featured dinosaurs as monsters always

---

**T. rex* is thought to have had a bite force of roughly 3 tons; by comparison, the largest great white sharks have a bite force of around 2 tons. Lions tend to have a maximum of 0.6 tons. Puny.

†Strictly speaking, they *are* still out there, they just have feathers and eat bread crumbs in your local park—but most folks don't think about pigeons and their kin as dinosaurs, even though, from an evolutionary perspective, they are.

latched on to the idea of a "lost world" where dinosaurs had somehow evaded extinction. During the years when Arthur Conan Doyle wrote *The Lost World* and Jules Verne penned *Journey to the Center of the Earth,* there were still vast swaths of uncharted jungle on the planet. The idea of a land where great beasts still reigned supreme seemed both plausible and exhilarating. These mysterious and unexplored landscapes were perfect for early science-fiction writers to make use of because even the top scientists of the age admitted they had no idea what they expected to find there.

Yet the "lost world" genre of dinosaur monster stories lost its edge as Earth got better explored and it became obvious that the only species going unnoticed were the insects beneath the feet of native tribespeople.

Michael Crichton's *Jurassic Park* dramatically changed the dinosaur monster genre by moving away from the historic "lost world" concept. Instead, Crichton, being well trained in the sciences, accepted dinosaurs as extinct and suggested that mankind had the ability to bring them back through modern genetics. Not only was this a new idea that the general public had not yet pondered, it also was one that researchers had not much considered either.

Resurrection attempted

The murky details of the science and theory behind *Jurassic Park* literally "resurrected" the beasts as monsters for a new generation. There were so many new unknowns in Crichton's story that imaginations ran wild with the possibility of a terrible fiasco, like that seen in the theme park featured in the book, actually happening. Questions ran so deep that, in the wake of the book and film, people began to ponder the possibility that what *Jurassic Park* suggested with regard to dinosaur resurrection was actually possible.

All it would take was DNA stored in a drop of blood. In *Jurassic Park,* the premise is that mosquitoes collected blood from dinosaurs during the Jurassic and Cretaceous Periods, and some of these insects then became stuck in tree resin. Tree resin becomes hard over time and transforms into amber, which, according to the book, protects the insect and the dinosaur blood in its belly from decay and destruction.

Jurassic Park researchers drill into amber samples and draw out dinosaur blood from the encased insect to put together dinosaur DNA. The concept captivated imaginations because, at the time, the science seemed feasible.

The Natural History Museum in London, having no shortage of dinosaur bones, a large collection of fossilized insects in amber, and a team of capable paleontologists, became a focal point for inquiries from both the research community and patrons smitten with the concept of bringing back the dead. A team was assembled to start looking for dinosaur blood inside fossilized insects and, if found, to determine whether any genetic information on dinosaurs could be collected from it.

Upon setting out on this exploration, there was a great deal of optimism. DNA falls apart over time. Any exposure to bacteria, dramatic changes in temperature, or shifts in pressure can cause the tiny sequences of proteins making up DNA to decay, leaving the genetic code in useless tatters. Yet the idea that an amber shell might afford protection presented an exciting possibility for preservation.

The process of tree resin hardening into amber causes objects stuck in the sap to dehydrate. This is significant because water removal is well known to shield organic materials from decay. This is one reason why properly prepared Egyptian mummies buried in the desiccating sands of North Africa are still in such great condition after more than three thousand years. It was this desiccating property of amber that researchers were counting on to help DNA survive the sixty-to-eighty-million-year time capsule journey from the age of the dinosaurs to the modern world.

To add to the enthusiasm, in 1992, just two years after *Jurassic Park* was written, a laboratory in California reported extracting insect DNA from an ancient bee that had been encased in amber millions of years ago. Shortly thereafter, reports started cropping up describing the recovery of amber-preserved termite and beetle DNA as well. With so many cases of insect DNA being recovered, hopes were further raised that dinosaur DNA from insect blood meals would soon follow.

To start their work, the Natural History Museum team tried to repeat the Californian insect DNA capture research. They attempted to collect insect DNA from the same extinct bees stuck in amber that the Californians had worked with but were unable to do it. Using multiple extraction methods and repeating these with several different specimens got the team nowhere. And this process was intended to be just the first step. Really old, and therefore highly valuable, amber samples from the days of the dinosaurs were to be drilled into and sampled only if insects in younger amber yielded helpful data. With no useful information coming from the process, nobody could justify continuing the research and the project ended.

In theory, the failure to resurrect dinosaurs using the methods presented in *Jurassic Park* should once again have made dinosaurs less viable as monsters in modern media. Yet with the release of numerous film sequels, dinosaurs maintained their monster status. A reason for this probably lies in the fact that *Jurassic Park* used a multilayered approach to make its monsters believable. The resurrection technique was the most obvious, but there were actually two more.

Nature untamed

Chaos theory isn't a big factor in the film, but the book develops the idea in great detail. Ian Malcolm, the mathematician in the story, relentlessly argues that biological systems, like those on display in the park, have tiny variables present in them that make their future behavior impossible to predict and, ultimately, impractical to control.

This talk of the natural world being unpredictable generates a perspective of nature that resonates with many real fears in society today. Hurricanes, tornadoes, earthquakes, and tsunamis wreak untold havoc and are extremely hard to predict. By their very nature they are chaotic and terrifying. Earthquake and tsunami detection research is advancing swiftly and, in the decades ahead, warnings of these disasters will become more common and precise. The same is true of extreme weather. Radar detection methods for predicting these dangerous events are improving, but it is still hard to work out exactly how hurricanes and tornadoes are going to behave.

Events like tornadoes and tsunamis, however, are not biological systems, and Malcolm's argument about chaos theory and the dinosaurs in *Jurassic Park* is that biological systems are uncontrollable by nature and therefore inherently threatening. Specifically, he says, "If there is one thing the history of evolution has taught us, it's that life will not be contained. Life breaks free, expands to new territories, and crashes through barriers, painfully, maybe even dangerously."

The chaos theory discussions are then combined with an image of dinosaurs not as extinct beasts with no real place on our planet but as living, breathing animals with a great many natural behaviors. The attention to ecological detail runs throughout both the book and the film. The most striking example is the moment when Dennis Nedry, the corrupt computer programmer who purposely shuts down the electric fences in the park to smuggle out stolen dinosaur embryos, gets attacked.

Nedry is killed by a dinosaur, yet the mechanism for the attack is remarkably similar to the attack behaviors seen in modern predators. In both the film and the novel, the dinosaur, which is described as belonging to the genus *Dilophosaurus,* chirps like a bird while spitting venom from a distance to blind its prey before closing in for the kill.

There are no animals that actually spit venom, but there are a number of cobras that spray venom out of their fangs. While these real "spitting" snakes can (and do) blind other animals with their venom, they do not spray their venom for hunting; they do it as a defensive measure when threatened by larger animals. And just like the venom shown in the film, the venom sprayed by cobras does cause searing pain and blindness if it strikes animals (or people) in the eyes.

Maintaining a distance while spitting is equally well founded upon reality. Many predators go to great lengths to stay away from their prey when they attack. Komodo dragons make a single venomous bite and then back away so the toxins can do the killing for them.* Venomous snakes do the same, allowing their venom time to take hold. The reason for these behaviors is thought to be that struggling prey can cause a lot of damage as they make their last attempts at survival. If a predator knows it has delivered a fatal or debilitating attack, there is no point in risking injury when dinner will be risk-free in just a short while.

The combination of such realistic traits in the dinosaurs along with the message that natural systems are totally unpredictable reinforces an image of the natural world as inherently dangerous to interfere with. In a sense, Jurassic Park makes Mother Nature the monster and the dinosaurs her tools for attack.

While snakes, as discussed throughout this book, certainly represent fears all on their own, a critical question to ask is whether there is any substance to Malcolm's arguments. Is the natural world really so chaotic and fierce that any attempt to intervene is going to lead to disaster?

Even though humanity has not played God on the scale that the geneticists do in *Jurassic Park,* there has been a lot of ecological meddling in the past hundred years that is not really much different. One example stands out.

Shortly after colonists arrived in Australia, sugarcane started being grown in the warm and wet northeastern corner of the country. The environment was perfect for the crop and yields would have been incredibly lucrative—if it hadn't been for the presence of a beetle. Unfortunately for the Australian farmers, the native cane beetle was fond of sugarcane. The adults of the species readily chewed up the plants while the larvae destroyed the roots.

*Komodo dragons were long thought to have toxic saliva that caused lethal raging infections in the animals they bit, but recent studies show that they have venom glands in their mouths.

In the early 1930s, the Australian government started considering the possibility of using nature against nature. Ideas were floated for bringing in a predator that could dramatically reduce cane beetle numbers and the government ultimately settled upon the cane toad from Hawaii, a carnivorous amphibian with a ravenous appetite for beetles.

The toads were introduced in 1935 and dismayed everyone when they were found to prefer insects other than cane beetles. Their taste for other insects was not a particularly large problem for the Australian environment, although it frustrated sugarcane farmers whose crops were still being destroyed by beetles. What proved disastrous was the fact that many Australian predators attacked the fat and slow-moving toads.

Cane toads have poison glands in their bodies, and unless they are eaten very carefully, their poison kills animals that consume them. Moreover, they were easy targets, which made matters worse since they were so readily hunted by Australia's unique (and often rare) predatory marsupials, birds of prey, and snakes. The predators ate the toads and died in growing numbers. Having no successful enemies, the toads have today taken over the northeastern corner of Australia and continue to expand their territory. Native predator populations in the region respond by dwindling. Australians hunt the toads aggressively and scientists are working on ways to interfere with toad fertility. A solution may arise in the decades ahead, but there is no denying that playing God and installing the toads in the first place has created an ecological nightmare and cost Australia dearly.

In the Grand Canyon, human meddling with the environment has caused similar damage. Dams have been on the Colorado River for nearly a century. However, there were no dams upstream of the Grand Canyon until 1963, when the Glen Canyon Dam was built. At the time, the idea of damming the river above the canyon didn't alarm anyone: Dams were popular, and cheap, clean hydroelectricity seemed like a good thing.

By the 1970s, serious long-term effects of the Glen Canyon Dam were becoming apparent. Downstream from the dam, the beaches that river rafters were used to camping on had eroded to half their original size and the National Park Service noticed that numerous native fish and animal species were vanishing.

Extensive analysis of the Glen Canyon Dam and many other dams around the world revealed that when dams slow water flow, they also disrupt the movement of sediment that the water is carrying. The principle behind this effect is relatively simple. When water moves slowly over bits of sediment on the river bottom, the force of the water might make the sediment roll along the bottom, but unless the sediment is very fine, like mud, it won't get picked up. When water moves quickly, however, the situation changes, and even relatively large grains of sediment can be carried away by turbulent water flow. In raging floods, small rocks and even boulders can be swept away.

Geologists realized that in the reservoir behind the dam, water was dropping just about all the sediment it was carrying. Since the water had been unmoving for quite some time, when it was ultimately released past turbines, it carried little sediment and came out the other side of the dam crystal clear. This was a problem for fish downstream that depended upon sandy coves and rocky habitats to hide in, but, like the cane toad situation in Australia, it was the unintended consequences of introducing a foreign species that proved disastrous.

Humpback chub, a rare species of fish unique to the Grand Canyon, were not deemed impressive enough to interest fishermen. To compensate for this, the Colorado River was seeded with trout eggs and a population was enthusiastically nurtured to support fishing tourism.

Trout are voracious clear-water predators that eat other fish. Humpback chub, in contrast, depend upon cloudy water to hide from danger. By building the dam and introducing new predators, humans made it easy for trout to spot and eat humpback chub. Today the chub survive in just a few of the small undammed streams feeding into the Colorado River. Whether they will survive in the wild for very much longer is anyone's guess, but they are now classified as just one step away from extinct.

To manage all of these problems associated with the Grand Canyon, the U.S. Geological Survey, in collaboration with numerous federal agencies, has taken something of a "nuclear" approach and, on several occasions, fully opened the Glen Canyon Dam floodgates immediately after heavy rains to bring sediment down cliffs and into streams to maximize the sediment's spread through the Grand Canyon. A lot of questions are being raised about the effectiveness of this approach, but the fact that such desperate measures are being taken hints at the severity of the problems created by a reprehensible lack of foresight.

With these two tales of outstandingly bad environmental management in mind, Ian Malcolm's argument of nature being an uncontrollable force seems accurate. Yet as severe as these two ecological catastrophes are, humanity has not always left the natural world mangled.

Real resurrection

Yellowstone National Park in northwestern Wyoming is home to a great diversity of animals. Visiting the park is very much like taking a safari, with a drive through the Yellowstone area rarely avoiding a traffic jam created by bison, elk, moose, or bighorn sheep that have no appreciation for roads or traffic.* Tourists often come away with the impression they have seen one of the few truly pristine wild places left in the world. Rather remarkably, they are wrong.

The story begins with pack rats, little rodents that often build their homes in rocky crevices. To construct their nests, they collect anything and everything from their habitat that piques their interest.

*Strictly speaking, this might not actually be true. A couple of recent studies suggest that large herbivorous animals are migrating toward roads to give birth specifically because they seem to realize the predators that eat their young (like grizzly bears) tend to avoid areas where humans are present in large numbers, like roads. Gives the term "human shield" a whole new meaning.

In the modern day, this unfortunately means they frequently pick up shards of glass, bottle caps, and other bits of junk, but long before humans made a mess of things, they collected feathers, plant material, and bone. The pack rats then brought back their findings and urinated on them. Bizarre as it sounds, their urine binds their odd collection together into something of a nest.

Paleontologists have long been interested in pack rat nests, or middens, as they are called, because pack rats are nonmigratory animals that stick with the same nest location for many generations. Each new pack rat brings in new collected material, dumps it on top of old material, and solidifies it into place by urinating all over everything. Indeed, in the past few decades, paleontologists realized that some middens could literally be thousands of years old and that they represented a vast store of preserved bones unwittingly collected by pack rats. They were like little paleontological gold mines.

In the early 1990s, as paleontologists led by Elizabeth Hadly at Stanford University were surging forward with their analysis of small bones found in Yellowstone middens, ecologists were asking serious questions about the park's ecosystem.

Elk seemed to exist in greater numbers in Yellowstone than in many other wilderness areas in northern North America. In addition to this, coyotes, which traditionally function as scavengers in most ecosystems, were found hunting larger animals than those they would usually pursue and were behaving more aggressively than coyotes in other regions. Some researchers suggested that this behavior was due to the absence of wolves, another large predator, which until 1923 had been living in Yellowstone but were wiped out by deliberate hunting.

Controversy raged over this issue. Many argued that wolves had artificially been in the Yellowstone area only during the late 1800s and early 1900s after being pushed out of lower-elevation areas where Europeans were expanding their settlements in the Great Plains. Yet Hadly's team disproved this theory.

Hadly found wolf bones that were thousands of years old in pack rat middens, proving that wolves had been in the Yellowstone area long before humans ever entered the picture. Her findings led ecologists to realize that wolves really were supposed to be living in the area; human activity had not artificially shoved them into the region. A recent local wolf extinction, caused by people, was the reason for coyotes functioning as predators and for the overly large populations of elk. With this in mind, the solution seemed simple: Reintroduce wolves to the park. Ultimately, to the chagrin of many ranchers and farmers in the area, they were returned to Yellowstone in 1995. Like *Jurassic Park,* animals driven to extinction had been brought back to the land they once roamed.

In the 1970s, Crichton's description of an uncontrollable, dangerous natural world was a reasonable conjecture. But scientists in the fields of ecology and conservation biology are getting better at making models of the systems they are thinking of tampering with. They also are studying past ecological activity by looking at fossils of the same animal over time so they can more accurately predict the effects of their proposed actions. This improvement in modeling is comforting, since global warming will require a great deal of human intervention if the world's diversity is going to remain intact as Earth continues to warm up. Indeed, governments are hiring paleontologists to analyze fossil ecosystems that existed during earlier times of global warming in order to work out how modern ecosystems are going to alter as climate changes.

In theory, as such environmental tinkering improves, the argument made by Ian Malcolm in *Jurassic Park* should weaken. This, in combination with the fact that dinosaurs cannot be resurrected using the methods outlined in the film, should make these creatures wane as monsters in popular culture during the decades ahead.

Yet resurrection of ancient beasts is not entirely out of the question. Hadly's team was able to find intact DNA in fossils thousands of years old in North and South American ecosystems. Hadly has no interest in even considering the resurrection of extinct beasts, but there are more than a few groups keen to make a lot of money by bringing back something like a woolly mammoth. The idea has led to interest from both the research community and the general public.

To some it just sounds exciting; to others it is a great challenge to be tackled. To pursue it, two avenues are being explored.

Some researchers are looking for germ cells, like sperm and eggs, in mammoths that were covered in ice in Siberia thousands of years ago. No research teams have succeeded in finding any such cells intact, but even if they were to do so, they would then need to use an elephant egg cell or sperm cell to partner with whichever mammoth germ cell they found to create a few "half mammoths." These half mammoths would then need to be intensively inbred with one another until a "near mammoth" that was only 1 percent elephant was created.

The other option for creating a mammoth involves collecting standard DNA from a frozen fossil specimen and injecting it into an elephant egg cell that has had its own DNA removed. This was the process used to create Dolly, the cloned sheep. No mammoth DNA of good enough quality has been found to make such a process viable, but researchers keep finding better and better samples, so there is a chance that such a project might one day prove successful.

Yet hope for resurrecting species is misguided. Not only is our planet in the exactly wrong climatic period to sustain these ice age beasts, vast resources are going to be required to make a project like this work, resources that could be used to save the many critically endangered animals on the planet. "How can we justify 'bringing back' animals when we are at a loss to preserve the diversity on the planet today?" asks Hadly.

A resurrected mammoth would inevitably end up in a zoo, since no natural habitat exists for such a creature. In contrast, if the carnivorous marsupial the Tasmanian devil goes extinct in the decades ahead—which is likely given that devil populations are dying out from a poorly understood cancer that gives them horrendous facial tumors—resurrection technology could much more sensibly be used to help their populations remain in their environment and survive the epidemic. There are even reasonable arguments for potentially resurrecting very recently extinct animals like the passenger pigeon and the Tasmanian tiger, another carnivorous marsupial that died out in the 1930s. These animals went extinct as a direct result of recent human activity, and their habitats are still more or less intact. Indeed, their habitats, similar to that of Yellowstone, have been thrown out of whack by their absence. From the perspective of maintaining ecological balance, pouring resources into resurrecting some animals makes sense to set things right again, but only after all available resources are expended to stem the tide of destruction that is wiping out ever more species. It is, after all, much easier (not to mention cheaper) to keep an extremely rare species alive than to attempt recovering one from beyond the grave.

With moral arguments standing strong against resurrecting even mammoths, and the possibility of ever using DNA to resurrect something as old as a dinosaur looking highly unlikely, will dinosaurs forever fade from popular view as monsters? This seems doubtful.

Where there's a will . . .

You can't keep a good scientist down, and among paleontologists this tends to be particularly true. There are numerous paleontologists who would love to see a dinosaur brought back to life, and even though the methods of resurrection presented in *Jurassic Park* have failed with dinosaurs, there may be another way.

Traditionally, when paleontologists have looked at the bodies of animals and asked, "Now how the heck did *that* evolve?" they have gone looking for the fossils of the given animal's ancestors. A classic example is the bird wing. Paleontologists know that birds evolved from animals that did not have wings initially, but the evolutionary pathway that was followed toward flight is one we do not understand very well. Did bird ancestors jump around in trees? Did they glide first and develop powered flight later? Or were feathers selected for by evolution simply because they looked sexy (to other birds) and meant that bird ancestors with more colorful feathers could mate more often?

Researchers have spent ages trying to solve the evolutionary puzzle of flight by searching extensively for fossils of very early birds and then analyzing the bones to better understand how ancient wings or, in some cases "proto-wings," would have worked. The work is truly astounding, and a lot of theories for how flight *might* have evolved have been put forth.*

Intriguingly, a research lab at McGill University in Montreal led by Hans Larsson is taking the study of great evolutionary transitions, like flight, in a rather new and extraordinary direction. The lab is intensively analyzing the way modern animals grow from their state of being a single cell, into masses of cells, and then on to embryos, fetuses, babies, and ultimately adults.

Why is studying babies useful when studying evolution? Are they not two totally different processes? Well, lurking in the genes of animals are shadows of their evolutionary past. These genetic shadows usually remain shadows as the animals develop from embryos to adults, but very rarely they can become quite real in the form of atavisms (remember those from way back in "Beastly Blends"?). Humans can sometimes be born with a tail and snakes and dolphins are sometimes born with the limbs that their ancestors had. What all this means is that many of the genes of the past are present in the body, but they just are not activated most of the time.

In theory, this means that instead of studying fossils to learn how a bird wing evolved, researchers can also study the embryos of birds, tweak their genes, and see for themselves which ones control the formation of wing bones from limbs. This is the sort of work Larsson and his team are doing and, as it happens, this is also the kind of work that may ultimately present another viable pathway for resurrecting dinosaurs.

---

*The theory that gliding led to powered flight looks to be utterly impossible. All early bird wings seem to have been fully mobile and flapping from the start, whereas modern gliding animals, like flying squirrels and flying snakes (yes, there are such things), just fling themselves outward from trees without moving their "wings" at all. Some nice work with bats is hinting that dropping down from branches followed by fluttering to help control the end of the drop might be how flapping flight actually evolved, but more work needs to be done.

By looking carefully at bird embryos and identifying the biochemical and genetic processes controlling certain parts of the bird's development, researchers are finding it possible to deactivate feathers from forming on certain parts of the bird's body, activate scale formation in certain areas, lengthen the tail, and, perhaps most dramatically, cause the bird to grow sharp teeth.

Now consider this: If a bird were to have its feathers removed, its skin covered with scales, its tail elongated, and its mouth littered with pointed teeth, what would we have?*

The number of genetic manipulations needed to actually create a dinosaur from a bird embryo is enormous, and it is going to be some time before anything viable is produced. But unlike finding usable dinosaur DNA in amber, there has not been much argument about whether the mechanisms for this procedure are sound.

Make no mistake, something with razor-sharp teeth is going to be hatched from a bird egg in the next hundred years. Let's just hope it doesn't get loose in the lab and start spitting venom.

*Okay, yes, a really ugly chicken, but you get the idea.

10

Extraterrestrial Threat—Aliens

"My mommy always said there were no monsters—no real ones—but there are."

—Rebecca "Newt" Jorden, *Aliens*

On the night of September 19, 1961, while driving south along Route 3 in New Hampshire, two people had a close encounter with something they described as resembling a large illuminated cigar-shaped object. They woke up later with no memory of the time that had passed and found signs on their clothes and car that led them to conclude they had been abducted and studied.

Whether this was the first reported alien abduction in human history or simply the first to get wide media attention is up for debate. What is not debatable is that these were fairly normal individuals. Neither had a history of being an alien enthusiast or showed signs of being mentally unhinged when they were later psychologically evaluated. For these reasons, their story was taken seriously and the location where the abduction is said to have occurred was commemorated by the state of New Hampshire with a plaque describing the event.

Countless tales of abductions and bizarre sightings followed.

Some of these reports came from people who were alien fanatics, but some were not. Yet there is something notable about the hysteria: All of it was taking place as the United States and the Soviet Union were racing into outer space. Just as the so-called space race was weighing heavily on the minds of millions, aliens were suddenly coming for a visit. The nature and timing of the alien abduction reports lend credence to the skeptics. Why were earlier abductions not reported to the press? If aliens can come to Earth with glowing cigar-shaped spacecraft, why does all the space-scanning technology that has been developed during the past decades never detect them? It is an enigma.

Regardless of all the questions, it is intriguing to note that alien life is now taken more seriously by the research community than ever before. Whether or not aliens are "somewhere out there" is a question that, scientifically, is not up for much argument anymore. In 1996, it was reported that evidence of microfossils of bacterial life were found in a Martian meteorite. The meteorite had once been part of the surface of Mars, and after the red planet was hit by another large meteor, it got blown off, and ultimately, after hurtling through space, found its way into Earth's orbit. Scientists found an odd mix of chemicals and minerals in the meteorite that looked a lot like chemicals and minerals sometimes left behind by simple microscopic organisms on Earth. This led the researchers to propose they had found evidence of life having once existed on Mars.

It might be easy to dismiss this work as the enthusiasm of some fringe researchers and perhaps a bit of lax publishing by a second- or third-tier academic journal with poor peer review processes, but the report was written by a team at NASA and published in the journal *Science*. This does not, of course, indicate that the findings were bulletproof. NASA makes mistakes and *Science* does sometimes publish articles that are later proved to be dead wrong. Dozens of articles followed the 1996 report, poking at potential flaws and questioning the scientists' interpretations. This is the nature of modern research, and it will take time before the greater scientific community concludes with certainty that these microfossils are actually evidence of life having once existed—or not—on Mars. Even so, scientific consensus is probably not as far off as it may seem. Evidence is mounting that liquid water was once present on Mars, indicating that conditions may have existed that could have allowed for the evolution of simple life. Furthermore, as astronomers explore the cosmos with advanced telescope technology, it is becoming apparent that there are other planets in distant locations of a similar size to Earth and a similar distance from a sun. This hints there are places where conditions much like those found on Earth can exist, and if this is true, the logic runs that life has probably taken root. It is on the question of what alien life is actually like where the wheels come off the proverbial wagon.

The shape of the unknown

In theory, if life were to evolve on a planet with conditions like those on Earth, it should follow evolutionary pathways similar to those that have happened here. Even so, some of evolution's pivotal moments have taken place after chance events. Volcanic eruptions, extreme glaciations, severe sea level changes, and meteorite impacts are just a few of the phenomena that have affected the course of evolution. This raises the fascinating question of whether life on Earth would be substantially different today if only a few major chance events took place at different times or did not take place at all.

As an example, consider the concept of snowball Earth. Millions of years before the rise of multicellular life, geologists theorize that the planet became extremely cold, with glaciers of enormous size covering the tropics. Some experts argue that the freezing was extreme enough to create ice sheets over the entire ocean. Others maintain that the cold was not quite so severe and that ice did not cover absolutely everything.* But most everyone agrees there were at least some glaciers in the tropics near the equator.

*A theory affectionately known as slushball Earth.

Glaciers look solid and unmoving, but they are actually rivers of ice flowing very slowly. As they move, they break pieces of sediment and rock from the land and carry these off. Many glaciers eventually flow onto the ocean, where they float out over the water. When they eventually melt, they drop the rocks they are carrying, which land among ocean sediments that are often distinctly different from them. Known as ice rafted debris, these "drop stones" provide crucial evidence of major glaciations having once taken place.

Patterns of ice rafted debris found in ocean sediments hint that there was not just one cold period, but many. The planet, for reasons that are still not well understood, seems to have gone from very warm to very cold numerous times just before multicellular life started to evolve. Some paleontologists argue that the timing of complex life's evolution closely following these climate oscillations was not a coincidence.

When life gets isolated into small and enclosed environments, selective pressures are altered. This is why birds that find their way to places without land predators often, over time, lose the ability to fly and why animals that end up on isolated islands look so different from their mainland kin. These evolutionary effects have led to the idea that extreme glaciations isolated simple life into tiny communities for long periods of time, allowed evolution to do some very strange things since selective pressures in these isolated environments were unique, and then released the organisms that evolved in these communities to interact and compete globally for a while before isolating them again in another extreme glaciation. This scenario may have functioned as a crucible of evolution and proved essential to the development of multicellular life.

But what if these extreme cold and hot episodes never happened? Would life have ever moved on from being single-celled? Would it have moved in a very different direction? The same sorts of questions are asked about the meteorite impact and other catastrophic events that struck the planet sixty-six million years ago. If *Tyrannosaurus* and *Triceratops* had not died out, would mammals have ever stepped into the evolutionary spotlight? Would reptiles still rule Earth?

For all of these reasons, alien life from a planet identical to Earth could be remarkably different from life as we know it today simply due to the fact that chance events would have likely shaped biology differently. Trying to work out what life was like on a planet with subtly different characteristics really makes things interesting. Consider a planet very much like Earth but with reduced gravity. Flying and jumping would probably evolve more easily, skeletons would not need to be as robust since weight would be reduced, and winds would throw much more debris into the air, possibly leading to the evolution of visual organs quite different from the eyes used on Earth. Gravity is just a minor change, but the evolutionary effects would be major. To put it simply, Earth-like planets are probably out there, they probably sustain life, and there is a good chance that any complex life found on them is vastly different from the life found on Earth. The mysteries that surround alien life are as vast as the blackness of space itself. It is this ponderous infinity from which terror takes shape.

Resistance is futile

One of the earliest and best-known tales of alien invasions is H. G. Wells's 1898 *The War of the Worlds,* in which aliens are technologically advanced and capable of launching a full-scale invasion of Earth from Mars. Although Wells's aliens are distinctly inhuman in form—they are large, oily, brown in color, and have blood-sucking tentacles around their mouths—they behave like aggressive human colonists, judging the technologically backward people of Earth as not worthy or capable of managing affairs on their own. For this reason, they attack the planet with the intention of using its resources for themselves.

Questions about the vastness of space and the unknowns associated with it are part of what makes *War of the Worlds* frightening. Certainly, when Wells was writing his masterpiece in England, many scientists in London were looking at the starry sky through telescopes and wondering about what might be found on the planets they could see. But crucial, frightening, and disturbing elements of the story actually have little to do with the monsters of this tale being aliens.

The late 1800s were a time of expansion of the British Empire, involving the ruthless colonization of foreign lands. Native peoples were treated terribly, natural resources were plundered, and countless people died, either directly at the hands of colonizing forces or indirectly from diseases carried by them. It is not hard to see the similarities between the aliens in *War of the Worlds* and the British colonists when the aliens are described as "intellects vast and cool and unsympathetic" that "regarded this planet with envious eyes." True, they lack the accent, bad teeth, pasty complexion, and fondness for tea, but they frequently disregard human life, are quick to enslave, have much better weapons than those they are invading, and often have an interest in taking resources or making "better use" of Earth than humans. The language that Wells applied to his aliens could be applied to the British colonizers. A popular theory suggests that Wells was not really writing about a fear of alien invasion but rather was offering a critique of colonization that forced everyone who read his story to reflect upon the horrors perpetrated by the British Empire and consider the possibility that one day such a fate could befall them as well.

One might think a fear of colonization could be appreciated only by people living during a time of imperialism, but this fear turns out to be more deeply ingrained in society than many people realize. Numerous movies, including several film adaptations of Wells's original text, have played with this frightening idea and done a good job of gripping audiences with it.

An example is Roland Emmerich's 1994 *Stargate,* in which aliens come to Earth during the days of the ancient Egyptians via a portal, rule over feeble humans like gods, and use them for slave labor. Also included in this genre are Emmerich's *Independence Day* and Tim Burton's farce *Mars Attacks!* The plot for both simply involves aliens coming to Earth and killing lots of people in an attempt to take over the planet. However, the concept of aggressive alien invasion is most chillingly presented in the form of the Borg species found in *Star Trek*.

The Borg live to force other species to take on their own form of perfection through enslavement. They violently invade and kidnap other races so they can infest them with cybernetic technology and later use these infested individuals as drones that follow the orders of the Borg collective. This abduction and transformation process is disturbing to watch, as characters lose their humanity and are forced to follow orders they would otherwise find repulsive. Ironically, the process is called "assimilation," in reference to the tactics used during the days of imperialism to make savages near colonies behave in a more "socially acceptable" manner.

Of predators and parasites

In spite of how frequently organized invasions are associated with extraterrestrials, there are many terrifying alien films that do not involve this subject at all. The best known is, of course, Ridley Scott's 1979 film *Alien*.

The crew of the mining vessel *Nostromo* check out a distress signal being sent from a ship stranded on an unexplored planet. Some of them drop down to take a look, and one person ends up wandering into a field of egglike structures. A creature bursts out of one of the egg sacs, breaks through the man's mask, and latches on to his face. The other crew members who are nearby drag this poor fellow back to the ship, where a conflict develops. The warrant officer, a woman named Ripley, insists that regulations be followed and the man be quarantined, but the others disagree and the crew member with the attached creature is brought on board. This turns out to be a big mistake.

All attempts to remove the alien from the crewman's face fail. It has acidic blood that sprays when it is cut and a tail tightly wrapped around the man's neck such that trying to rip it off would result in strangulation. Eventually the alien releases its hold and dies. The man seems fine, but soon he doubles over in pain, and an alien, which has been incubating inside him, bursts out of his chest. It runs off and spends the rest of the film picking off the crew one by one with its sharp claws and teeth.

At its most basic level, the terror of this story is hardly new. The alien almost always attacks from dark alcoves, ventilation ducts, and tunnels. It emerges silently and unseen until it is too late for its victims to escape. This is identical to the behavior of many modern predators. Numerous great cats sneak up on mammals that they are keen to eat. The same goes for a number of sharks, including great whites, that sit in the dark depths below their surface-swimming prey and lunge forward to attack only when they think they have the element of surprise. The monster in *Alien* is playing off of the ancient fear of predators that sits deep within the mind of every human being. Other films, like John McTiernan's 1987 *Predator* and Nimród Anatal's 2010 *Predators,* which both feature an alien species that has a love for hunting humans, prove it is a theme that still thrills viewers, but it would be oversimplifying matters to entirely dismiss Scott's monster as the Nemean lion reborn.

The life cycle of the alien in Scott's film is a remarkably well thought out element of the story that is frightening all on its own. The alien species lays eggs that hatch into face-sucking larvae. These larvae then insert embryos into humans, ultimately transforming into adults that kill their hosts as they emerge. To modern biologists, this sort of a life cycle is well known.

The malaria parasite spends most of its time living inside humans, sucking up nutrients and causing a lot of harm. However, it cannot, all on its own, jump from one human to another. To spread its offspring to others, the parasite must allow itself to be collected by a mosquito feeding on human blood. Then, when the mosquito bites another human, the parasite offloads and begins a new infection.

Along a similar vein, schistosomiasis, which plagues much of Africa, is caused by a parasite that lives part of its life inside humans but spreads its offspring into water sources by traveling out of the body in human waste. Once in water, the offspring of the parasite infect snails, where they grow and mature. Ultimately, they swim out of these intermediate hosts and travel to bare human feet in nearby water. They burrow through the flesh of the foot, enter the body, and begin a new infection.

Perhaps the parasite that most closely resembles the monster in *Alien* is the human botfly of Central America. It grabs female mosquitoes and attaches its eggs onto their bodies. When these mosquitoes later feed on blood, the eggs hatch, the larvae drop down onto the flesh of the animals being fed upon, and begin burrowing. Similar to *Alien*, they mature into adults inside the human body and later burst out in a bloody and painful mess. It is utterly gross. And while this all sounds bad, these parasites are nothing compared to the parasites that other species must put up with.

Work being done by the parasitologist David Hughes at Pennsylvania State University reveals that ants in the tropics are often attacked by a group of fungi belonging to the *Ophiocordyceps* genus that stick to the insects' exoskeletons, inject themselves inside their bodies, and mess with their minds by releasing chemicals into their brains. These chemicals lead the ants to seek out locations where other ants are often found. Once there, the infected ants climb nearby plants and walk out onto the undersides of leaves. Here, they bite the leaves with their jaws and attach themselves firmly in place just as the fungus kills them by consuming their brains. The fungus quickly spreads through the bodies of the ants, consuming all it can. It uses the nutrients it has sucked up to build a reproductive structure that emerges from the bodies of the insects. This organ releases spores that rain down on all of the other ants passing by below. These spores stick to their bodies, infect them, and start the cycle all over again.*

It is entirely logical that Scott's 1979 monster parasite functions

*Intriguing new research by the parasitologists at Penn State is showing that malaria parasites can tinker with the minds of mosquitoes, making them more thirsty for human blood than the blood of other animals. This, of course, suits the human malaria parasite just fine since it cannot reproduce inside the bodies of other species.

much like many real parasites alive today. Since nobody has any idea what alien life would look like, it makes perfect sense for the creators of an alien monster simply to turn to real-world horrors as they conjure up a terrifying creature. Yet it is curious that such a monster does not grace the screen until 1979.

The subtle lives of multiple host parasites have been well known, and properly dreaded, since the days of Victorian specimen collecting. The Natural History Museum in London, where many of the most prestigious Victorian collectors sent the strange specimens they found in remote parts of the world, is loaded with the preserved organisms that were used to study how various parasitizing worms and insects lived.

It was during the 1880s when doctors first realized that malaria was caused by a parasite and related to mosquito bites; before this, nobody knew why so many people caught the disease. Although parasites were not popular movie monsters until Scott's parasite-like alien stepped into the limelight, they did make an appearance in early horror literature.

In 1939, Alfred Elton van Vogt published a short science-fiction story titled "Black Destroyer." This thrilling tale features the crew of a spaceship who are attacked by an alien that comes aboard and has the ability to insert its eggs inside human bodies. These eggs incubate, hatch, and kill their hosts in a manner startlingly similar to *Alien,* hinting that human fear of parasites was already alive and well back in the 1930s.

The sequel to *Alien,* James Cameron's 1986 *Aliens,* further played off fears of parasites by revealing the existence of an egg-laying queen and drones that collect human hosts for infection.* The television

*Technically, *Aliens* raises a challenging evolutionary question that borders on being a story flaw. It is not in the best interest of any parasite to kill its host outright since it needs the host as a living environment for its young to develop. This is presumably why the aliens capture the little girl, Newt, toward the end of the film and stick her in a cocoon rather than just killing her. Yet lots of marines are violently ripped to pieces by the aliens instead of being captured. Since every dead human results in one fewer alien ultimately being series *X-Files* also made good use of these same fears. In the episode "Ice," which appeared in the series' first season, arctic worms use humans as hosts. And in "Firewalker," a fungus-like organism found deep within a volcano treats humans in exactly the same way as the *Cordyceps* fungi treat ants, controlling their minds and infecting others with spores sprayed by stalks that burst out of human bodies in a rather disgusting manner.

X-Files explored the range of what an "alien" could be defined as, but it mainly took on predatory and parasitic forms mirroring both ancient fears of carnivores and more recent fears of horrific parasites. This is all entirely understandable. While few people today have a credible reason to fear being eaten by a lion, human evolution has likely hardwired into the brain a fear of being preyed upon, giving modern storytellers an easy terror access point. As for parasites, new species, including many closely related to the *Cordyceps* genus, are constantly being found and the powers of parasites over people are only just being discovered.

The protozoan *Toxoplasma gondii,* which spends much of its life inside felines, jumps from one cat to another by infecting mice via cat waste. Inside these rodents, the parasite interferes with their brains, causing mice to lose their natural fear of cats and triggering them to be attracted to their scent. As a result, infected mice become easy prey. This is perfect for the parasite that needs to end up in a feline body to complete its life cycle. However, this does not solely take place in cats and mice. Toxoplasmosis also has the ability to infect people, and if an infected human is eaten by a great cat, researchers speculate that the parasite can complete its life cycle just as easily.* And this is where things get creepy.

---

born, such behavior presents an evolutionary quandary. It makes no sense for the aliens to be such capable human-killing machines. Instead, they should be masterful human kidnappers that are adept at feeding on some other species when they reach adulthood. Given the success of the film, it seems likely that nobody cared, but I noticed.

*Getting permits to infect humans with toxoplasmosis and then feed them to lions in a lab setting is, understandably, an ethically tricky matter.

Recent research shows that toxoplasmosis has a lot of effects on human behavior, one of the most fascinating being impaired hand-eye coordination. Worryingly, one study found that infected humans have increased chances of being in automobile accidents. So does this mean the parasite has a history of hampering human physical coordination so lions and tigers could more easily catch ancient people and become infected while feasting on human flesh? It is too early to be certain, but work in this area is swiftly moving forward and the horrors of toxoplasmosis and many other parasites will continue to make it into the news in the decades ahead. With this in mind, it seems likely that aliens and other monsters that carry parasitic attributes will continue to be rather common. Yet in spite of the many horrific forms that aliens take, there is no denying the fact that not all are bad.

Alien innocence

It is easy to tremble at the vast inky blackness of space, but it is just as easy to look up at the night sky, marvel at the beauty of the stars, and make a wish. Before astronomers discovered the other planets in our solar system, the concept of aliens was, quite literally, alien. This makes the shadowy veil of space distinctly different from the shadowy veil that was once created by jungles and the oceans. Unlike these two other dark, monster-inspiring locations, space was often viewed with wonder before it was viewed with fear. It is this dual nature of space from which aliens of a good nature are also born.

Robert Wise's 1951 film *The Day the Earth Stood Still* initially runs along a vein similar to *War of the Worlds*. In the film, aliens that are far more technologically advanced than humans arrive on Earth and present a threat by their raw power—the lead alien's assistant is a robot capable of shooting out powerful lasers from its eyes. However, while the aliens in *War of the Worlds* immediately attack, Klaatu, the alien in Wise's film, tries to speak with the leaders of Earth to warn them that their violent ways are a problem.

Intelligent alien species living "nearby" in space are concerned about the human invention of nuclear weapons and human tendencies to declare war on one another. Klaatu explains humans must change their behavior or risk being destroyed. While Klaatu is not soft and cuddly, his overall appearance and relationship with humans are very different from aliens that engage in invasion, predation, and parasitization. And he is hardly alone.

In Steven Spielberg's *Close Encounters of the Third Kind,* aliens collect people, presumably for study. They are initially presented as frightening even though they are never actually seen, due to the brightness of the lights on their ships. Really, the fear invoked during the first three-quarters of the film is based on not much being known about these creatures. However, this fear vanishes once it becomes clear the aliens simply wish for peaceful contact and look like pudgy cherubs. In a sense, the film plays with the very essence of what it is to be an alien monster by pointing out that it is human perception of the unknown that brings on fear rather than any sort of violence spread by alien beings.

In J. J. Abrams's 2011 film *Super 8*, an alien goes on a rampage, abducting and killing humans in a small town. For most of the story, the alien seems like a classic evil monster. When it is ultimately seen toward the end, it is huge, insectlike, and clawed, much like Ridley Scott's beast. It attacks and kills many people, but as the unarmed child protagonists engage with the creature, they learn that it is humans who have done wrong. After landing on Earth long ago, the alien was captured by the U.S. military so it could study the technology of the alien's spacecraft. The children learn that the creature is lashing out only because it is hungry, scared, and longing to go home. Of course, when it comes to aliens feeling trapped on Earth and wishing to go home, there is one film that set the standard for all others.

In 1982, Steven Spielberg gave us an alien that was as different from Scott's parasite as possible. Small, innocent, and lovable, E.T. just wants to get back to his own world. But as different as *E.T.* is from so many other alien tales, there is one element that arises in all of these stories with remarkable consistency.

In *E.T.*, the antagonists are adult humans who want to capture E.T. for study. They are portrayed as uncaring about the actual welfare of the likable little alien, creating a striking difference from the children of the film, who view E.T. with a sense of wonder and love. This element is also present in *Super 8*, where the adults seem to have little interest in the plight of the alien and only the children are capable of seeing the harm being inflicted by humanity. Neil Blomkamp's 2009 film *District 9* presents the same sort of story, with humans horribly treating helpless aliens who have come to Earth as refugees. In one particularly heartless scene, the protagonist laughs at the popping sounds made as alien embryos are set on fire; it is cringeworthy stuff. And unlike Abrams and Spielberg, Blomkamp inserts no innocent children into his tale to give audiences any sense of hope for the future.

Remarkably, this story element is found in both *Alien* and *Aliens*. Make no mistake, the creatures in these films are definitely monsters that have no redeeming features whatsoever, but villainous humans play a role in both. In *Alien*, the villain comes in the form of a faceless corporation eager to capture an alien, regardless of the cost to human life, so the creature can be harnessed and used in military science. In *Aliens*, the villain is a slimy corporate executive keen to smuggle aliens back inside the bodies of infected humans so he can sell the alien embryos on the black market.

In essence, the villainous trait that adults carry in recent alien films is a willingness to be cool and unsympathetic toward life. "Cool" and "unsympathetic" . . . the words used by Wells to describe his aliens back in 1898. Have we somehow become the monsters?

In E.T., the antagonists are adult humans who want to capture E.T. for study. They are portrayed as menacing throughout the length of the feature film, again creating a striking difference from the children of the film, who view E.T. with a sense of wonder and love. This element is also present in Super 8, where the adults seem to have little interest in the plight of the creature and only the children are capable of seeing the horror being inflicted by humanity. Neill Blomkamp's 2009 film District 9 presents the same sort of story, with humans horribly treating helpless aliens who have come to Earth as refugees. In one particularly heartless scene, the protagonist laughs at the popping sounds made as alien embryos are set on fire; it is cringeworthy stuff. And unlike Abrams and Spielberg, Blomkamp inserts no innocent children into his tale to give the audience any sense of hope for the future.

Remarkably, this same element is found in both Alien and Aliens. Make no mistake, the creatures in these films are definitely monsters that have no redeeming features whatsoever, but villainous humans play a role in both too. In Alien, the villain comes in the form of a faceless corporation eager to obtain the alien, regardless of the cost to human life, so the creature can be harnessed as a weapon in military science. In Aliens, the villain is a slimy corporate executive keen to smuggle the alien inside the bodies of infected humans so he can sell the alien embryos on the black market.

In essence, the villainous trait that nearly every recent alien film has is a callousness to the weird and the unexpected. "Cool" and "unsympathetic" are the words used by Wells to describe his aliens back in 1898. Have we somehow become the monsters?

Conclusion

Cool and unsympathetic human behavior is not unique to alien films. While giant animals, like the Nemean lion and Calydonian boar, were celestially created to plague humanity, Kong, in every version of *King Kong,* is always brought to civilized lands by people who just don't care about the needs of the giant ape. Similarly, Chimera was a divinely spawned horror, but the chimeric Dren from *Splice,* Caesar from *Rise of the Planet of the Apes,* and Frankenstein's monster turn toward violence largely because the human world abuses them. Like Kong, they are viewed as specimens rather than as individuals, and this creates the conflict.

This is not to say that these mistreated monsters are not still monsters. They are aberrations and they do harm humans, so they are, by definition, monsters, but their motivations make them more complex and hint at a trend.

The ancient monsters were created by gods. True, they were sometimes sent to punish humans for misbehavior, as was the case with the Minotaur and Calydonian boar, but this was still often the result of the gods being greedy for attention, uncompromising, and harsh. During the Middle Ages, monsters remained largely disconnected from humanity. Creatures like the Rukh and dragons plagued the world, but this was not because humanity had brought them about; they were simply there. However, there is a subtle difference between these monsters and those of the ancient period. Even though people might not have brought about creatures like the Rukh or dragons, humans definitely elicited the Rukh's anger by trying to steal its egg, and dragons were always presented as attacking because people were trying to steal their treasure or attempting to enter their living spaces. Thus, humans took a more active role in interacting with monsters by overstepping boundaries.

In the late Middle Ages and Renaissance, the relationship between humans and monsters changed yet again. With the rise of demons that could possess vulnerable sleeping minds and vampires that could infect with a bite, humans started to become monsters. Yes, the humans who were corrupted by demons, vampires, and werewolves did not enter their state of corruption willingly, but this was still a major change from the way humans and monsters had been interacting earlier.

During the industrial and modern period, humans consistently have been side by side with monsters. While Dr. Frankenstein's monster, the vicious *Velociraptors* in *Jurassic Park,* the virus in *Contagion,* and the shark in *Jaws* are all terrifying in their own right, human behavior in these cases is largely responsible for these creatures getting utterly out of control and killing many more people than they otherwise would. Humans are not actually monsters in these stories because they do not fit the definition, but in recent years this has started to change.

I see you

In 2009, James Cameron's *Avatar* drew audiences of epic proportions. Set many years in the future, the story focuses on a mining operation taking place on the distant moon Pandora. The humans running the show are all corporate types, eager to get at Pandora's vast mineral wealth. Working with them are two other groups, ex-marines keen to earn more cash than they could fighting wars on Earth and scientists who are supposed to be studying the biology of Pandora as well as easing tense relations with the humanoid alien population. These aliens, the Na'vi, are blue giants with yellow eyes, leopard-like stripes, fangs, and tails—aberrations in every sense of the word. Unfortunately for the managers of the mining operations, the Na'vi have their home right on top of a very dense cluster of mineral wealth. They are reluctant to slaughter the aliens because "killing the indigenous looks bad." Thus, the scientists are ordered to intervene. Using technology to connect their minds to the bodies of avatars grown from a mixture of Na'vi and human DNA, the scientists are sent to interact with the natives and convince them to move.

The lead character in the story comes in the unlikely form of Jake Sully, a paraplegic marine who, through unusual circumstances, gets assigned to the avatar team being managed by the scientists. His initial experience on Pandora is dreadful. After being briefed by the military commander working for the corporation that "out there beyond that fence every living thing that crawls, flies, or squats in the mud wants to kill you and eat your eyes for jujubes," he is viciously attacked by nocturnal six-legged wolves,* nearly trampled by a hammer-headed rhinoceros-like creature, and almost eaten by a sleek black lionlike beast.

The night scene where he is attacked by the viper-wolves is utterly typical of monster movies. The camera initially reveals only the predators' glowing eyes fleetingly as they close in for the kill. Then, as they approach, the camera moves as if mounted on their backs. The protagonist's early interactions with the Na'vi are similar. They hiss, show their fangs, and appear truly alien.

Yet the world of Pandora is flipped upside down for both Jake and the audience as the story progresses. It becomes apparent that the land and the animals are deeply interconnected, and the aggression he met early on was all a matter of misunderstanding the ecology of the planet. This is most strikingly presented when Jake, while being shown the wonders of the forest by the Na'vi, finds the viper-wolves nursing their young and playing in their den.

But perceptions of the animals are not the only ones that change.

---

*Cleverly named "viper-wolves." Drawing upon some snake fears, are we Mr. Cameron?

By constantly watching Jake take action in a Na'vi body, we, as an audience, have our perceptions of who and what the heroes of the story are reversed. Halfway through the film, we begin seeing the Na'vi as normal and the humans as outsiders; even Jake pauses at one point to consider this, commenting, "Everything is backwards now, like out there is the true world, and in here is the dream." This is all helped along by subtle shifts in language as the characters begin describing the humans as "aliens." But all subtlety comes to an end as the tale concludes.

Frustrated by the constant problems presented by the Na'vi and tired of waiting for them to be coaxed out of the way, the corporation wages war. Humans, now frequently encased inside robotic walker units that make them look distinctly nonhuman, attack. Hundreds of Na'vi are mercilessly killed by missiles and machine guns. Finally, revealed to be somewhat sentient, the moon Pandora herself sends bestial hordes into battle. The viper-wolves, the hammer-headed rhinoceros creatures, the lionlike beasts, and the dragons of the sky come pouring out of the wild to kill the humans. It is utterly impossible to not favor these "once monsters" as they trample and rip people apart. By the close of the film, roles are entirely reversed. The aliens are now honorable humans, the monsters are now animals in need of protection, and the humans are now violent technologically equipped horrors much like the Terminators of Cameron's earlier films.

While many other movies and books have presented tales of tribes that are initially viewed as threatening and later found to be heroic (*Dances with Wolves* and *The Last Samurai* are good examples), what Cameron manages with *Avatar* is something of a first—he creates this same reversal with creatures that are utterly inhuman at the start.

Part of the film's success in so completely reversing the roles comes from the performance capture technology that enabled actors to perform with remarkable believability as aliens. All actors who played Na'vi or avatar characters were outfitted with suits that tracked their movements and wore cameras that mapped their facial expressions onto the faces of the fictional characters they were playing. It was this technology—which was, incidentally, first developed for the both loved and hated creature Gollum in Peter Jackson's *Lord of the Rings* films—that made it possible for the humanity of the actors to ultimately shine through the bodies of the aberrations they were playing and allow viewers to think of them as human.

Yet film technology is only partially responsible for allowing the monsters and humans to so completely swap places in *Avatar*. Another tactic that Cameron uses to make this reversal involves scientific methods.

A key Na'vi greeting is "I see you," which the film explains means "I see into you" or "I understand you." The Na'vi do not use this phrase in a scientific way, but such communication primes us to see Pandora differently. Like biologists looking below surface tissues with X-rays or geologists drilling through the Earth's crust for mineral traces, both Jake and the audience are taken behind the curtain of the forest and shown the ecological realities of Pandora that the corporate and militaristic humans simply cannot see.

Jake comes to see the creatures as audiences have long seen Kong, with understanding and sympathy, and he realizes who the monsters really are. In many ways the effect of this understanding is similar to what drove so many ancient monsters to extinction. Just as the Nemean lion met its end when the wilds of Europe were explored and what had previously been considered a monster came to be seen as merely animal, the monsters of Pandora are transformed as Jake "sees into" them.

So Cameron manages a remarkable thing with *Avatar*, but like any scientist making a major discovery, he is standing on the shoulders of giants who have been moving in this direction for a long time. Charles Darwin and Alfred Russel Wallace came up with the theory of evolution through natural selection independently and simultaneously because all the pieces of the puzzle were present for them to take this major step. Cameron was similarly primed by numerous films, many of which he directed, that were already bringing humans and monsters into close proximity. A monster/human reversal has been a long time coming. The critical question is, why? Why, throughout history, have humans and monsters grown ever closer together and finally, in *Avatar,* entirely swapped positions?

Where the mask finally falls

People have always looked to the horizon and feared that which they did not understand. Initially, this horizon was the edge of the forest. Then, when forests became better explored and their dangers were realized as not actually being that serious, human attention turned toward the darkness of the sea. Then the sea became better explored, and the new horizon became the vastness of space. And now, with space getting ever better explored, a new horizon appears . . . in the form of the horrors humanity is about to unleash on itself.

Through the technology we have created, we now pose a greater threat to ourselves and our planet than we ever have before. During the days of the Romans, the worst humanity could manage was to wage a war of swords. True, this frequently left thousands dead, but the world and all of its resources would always recover because nobody, not even the merciless Genghis Khan, had the power to exterminate everyone. Nuclear weapons radically changed this by making it possible for us to literally bring about the end of the world. Similarly, we have the ability, through reckless inaction and greed, to wreak environmental havoc that can make our planet truly unlivable.

Trying to predict what we will and won't do in the years ahead is as difficult as trying to determine whether the animal on the other side of a jungle thicket is a rabbit, lion, or dragon. It is from the nature of these threats and the staggering uncertainty surrounding them that we rise as monsters.

This should not, for a moment, indicate that the days of inhuman monsters are over. Nothing can alter the ways evolution has shaped the human brain, and people will always have an intrinsic fear of things that have threatened them since the days of dwelling in caves. Dark environments where we were once predated upon will continue to generate a sense of dread, and animals like snakes will remain objects of fear. Monsters associated with these stimuli will certainly persist for decades regardless of the role that we come to play in monster stories. Similarly, forces that still threaten human lives and are difficult or impossible to control, like diseases, will continue to find their way into the media as monsters. Even so, there is no getting around the fact that the mask of the monster has fallen over time with ever greater frequency on the face of humanity.

Whether this trend will persist depends greatly upon how we act. We can stand petrified as we gaze at the monsters we have become and allow worldwide nuclear and environmental destruction. Alternatively, we have the opportunity to take action, behead the beast, and claim a future where the mask of the monster safely sits somewhere else.

Afterword

Given the way arguments are made in this book, it probably will not come as a galloping shock to you that, before becoming a science journalist and author, I studied as a paleontologist.* Like many young paleontologists, I was initially drawn to the field by dinosaurs, but what ultimately got me hooked was the storytelling involved in explaining how life in every given period of history came from the life before it.

As a paleontologist you learn that nothing, absolutely nothing, in the history of life "just appears" out of nowhere. The earliest amphibians came from fish, the earliest reptiles came from amphibians, birds from reptiles, and the paleontological storytelling explains how these transitions took place. Sometimes the forces on our planet drove new species to appear quite quickly, sometimes the changes happened slowly, but sudden conjuring of species from thin air, that is not ever part of the story. Even the very earliest and simplest life-forms came from organic compounds present on our planet, and those came from materials that appeared when the Earth was formed.†

It was with this paleontological mind-set that I wrote *The Science*

*Yes, like Ross from *Friends*. Oh, if only I had a dollar for every time that was said when I was in university.

†And the only place where you can even start to speculate about that is with the big bang, but since I am hardly a Sheldon Cooper, I won't go there.

of Monsters, and it was not until I was touring the book that I ran into the simplest but most vexing of questions. I had just finished giving a lighthearted talk on the science of zombies to a group of psychologists at Northwestern when a young fellow walked up to me and asked, "Did you ever consider the possibility that the Haitians were just making zombies up to scare their children into doing their chores around the house?" I was jet-lagged, hungry, and tired, and batted away the question by pointing out that the work done by Wade Davis on zombie brews was top-notch and that the evidence of modern-age abductions by zombie masters was also pretty strong. There was no way zombies were just made up, end of story. However, a week later the question popped up again when another student who had listened to one of my talks ran into me at the airport.

"How do we know if these monsters really came from observations of the natural world? Couldn't people have just imagined them?" she asked. Having nowhere to go (I was just sitting there with my coffee waiting for my flight) and having had a good night of sleep, I gave her question my full attention. "The short answer is, We don't really know," I said. "There is no way for us to be sure what our ancestors were actually thinking. All we can do is make educated guesses." She asked how we make those guesses, and what followed was a long chat between us about paleontology, where I explained that we live in a world where nothing new just spontaneously appears out of the ether. Everything comes from something else. She nodded politely as I stood up to leave for my flight. Then, as I was walking away, she asked me one last question. "You are assuming our ideas are like animals and evolve as they do. Are you really sure that's true?" I said "Pretty sure," but, in honesty, I had not given it much thought.

So this raises the question, "Could we have just conjured the monsters of our past from our imaginations?" Yes, of course we could have, but the evidence suggests not.

Think about it . . . just how many ideas in our world are genuinely new? I went to see *Man of Steel* last night with a friend, and while it was fun to watch, the story is anything but new. The planet Krypton is dying because Superman's alien race has mined it too extensively.

Sound anything like the environmental challenges we face here on Earth? Superman's parents had sent him off from their dying world in a celestial basket to be raised in a far-off land. Exodus and Moses come to mind? Superman himself is warned by his adopted father that he should not reveal his powers because humanity is not ready for him. Jesus Christ anyone? And to top it off, we must remember that this all comes from a comic book published ages ago! In fact, the vast majority of our movies are rehashed versions of old ideas. In my conclusion to this book, I wax lyrical about *Avatar,* and while I do not deny that I enjoyed the film, it was a very straightforward rehash of *Dances with Wolves* and *The Last Samurai*.

With that said, among all of the rehashed content in *Man of Steel* (admittedly rehashed in a wonderful way), the one truly new touch that I appreciated was the idea that the "S" on Superman's suit was not actually a letter of our alphabet but a symbol in his alien language meaning "hope." I am not a comic book guru, but I am fairly certain that this was a novel idea. But was it actually new or was it just a modification of an old idea? And what actually constitutes a new idea anyway?

Ideas evolve from one another. Fiction constantly works this way with stories and characters subtly shifted around and tweaked as repackaged old ideas. Even science works in this manner with new discoveries based upon the discoveries that came before them. Indeed, we look back at Charles Darwin's argument that evolution was the result of natural selection as a wonderful bit of inspiration, but we are quick to forget one very important thing: Alfred Russel Wallace came up with the exact same argument at the same time, independently. So was it true inspiration that paved the way for the theory of evolution via natural selection to emerge, or was it a matter of the scientific, cultural, and philosophical climate being right for the idea to take shape when it did from the ideas that had come before it? I would argue a bit of both and suspect that the ideas underpinning monsters are not much different.

The Calydonian boar, the Rukh and the Nemean lion all use animals from the real world as their basic structures. Yes, the Calydonian boar and Rukh are larger than their real-world counterparts and the Nemean lion is more robust, but the changes are minor. Moreover, we know, without any doubt, that boars, lions, and eagles were common in the world around the people who came up with these monsters. So I ask, are these monsters the stuff of pure imagination? Or are they, like most of the elements in *Man of Steel*, rehashed versions of other ideas and experiences that had already been around for a long while?

An argument could be made, and it has been by a handful of aging historians, that oddball monsters like Chimera and Medusa are impossible to consider as anything other than purely imagined. They point out that the combinations of traits associated with these monsters are so completely unnatural that the only way to possibly explain them is by throwing up our arms and saying "It was just fiction." I disagree. What such stodgy classicists* fail to realize is that, strange as the mixtures of monstrous traits might be, they are still highly reflective of the natural world and the dangers that it posed to people.

The concept of Chimera breathing flames may or may not have stemmed from the eternal fires of Chimera that still burn today in Turkey, but there is no question that the fear of being burned alive is one that people have had since our earliest days living on the plains of Africa, where flash fires spawned by lightning strikes were all too common. Such fire fear may simply have been added to the Chimera to make an already frightening monster even more terrifying, or it may actually have stemmed from the discovery of blackened bones pulled out of the ground. We will never know, but what we cannot deny is that the creation of the Chimera's fiery breath stems from a real danger that people were truly afraid of.

Medusa is no different. Her petrifying gaze may or may not have had some connection with terrifying snake interactions that caused

---

*Who spend more time thinking about Alexander the Great than is probably healthy for them.

people to become paralyzed with fear. Alternatively, her gaze might have come from the finding of petrified human remains inside caves. Again, we cannot know for certain. However, what we *do* know is that paralysis through both fear and petrification are natural phenomena that our ancestors were grappling with.

More modern monsters make the point even clearer. The laboratory-born dinosaurs of *Jurassic Park* were not imagined from nothing; they were born from deep concerns about what the field of genetics was capable of long before the book was written.* Heck, even Frankenstein's monster did not come into being when it did at random. The scientific, cultural, and philosophical climate of the time presented people with the stimuli that they needed to conjure a monster. Yes, Mary Shelley gave public fears of transfusions and transplants physical form, but the act of creation that she engaged in was guided by the anxieties of the time. This does not make *Frankenstein* any less impressive as a story, but it was not just imagined out of thin air. Metaphorically, the monster's bits were lying out there to be pieced together before the book was written, and if Shelley had not put them together, someone else would have.

So, do we just make our monsters up? No, we don't. At least not any more than we make anything else up. We are masters when it comes to using information that we glean from our environment in new and interesting ways. We adapt, repackage, and stand upon the shoulders of those who came before us to tell fantastic stories, write impressive scientific theorems, build wonderful things, and yes, even create monsters.

---

*Let's not forget that dinosaurs did actually walk the earth . . . so the physical forms of the beasts in the film were definitely not just conjured up either.

Sources

Abbott, Alison. "Regulations Proposed for Animal-Human Chimaeras." *Nature* 475 (2011).

Angaran, P., et al. "Syncope." *Neurologic Clinics* 29, no. 4 (November 2011): 903–25.

Angras, W. S., et al. "The Epidemiology of Common Fears and Phobias." *Comprehensive Psychiatry* 10 (1969): 151–56.

Apollodorus. *The Library.* Translated by Sir James George Frazer. New York: Putnam, 1921.

Apollonius of Rhodius. *Argonautica.* Translated by R. C. Seaton. New York: Putnam, 1921.

Arrowsmith, Aaron. "Chart of the Dangers in the Channel Between Sardinia, Sicily and Africa." 1810.

Artemidorus. *The Interpretation of Dreams.* Translated by Robert J. White. Park Ridge, NJ: Noyes Press, 1975.

Asimov, Isaac. *I, Robot.* New York: Gnome Books, 1950.

Aymar, Brandt, ed. *Treasury of Snake Lore.* New York: Greenberg, 1956.

Babes, Victor. *Traité de la Rage.* Paris: Baillière, 1912, pp. 97–103, 439–40.

Bailey, Geoff, and John Parkington. *The Archaeology of Prehistoric Coastlines.* New York: Cambridge University Press, 1988.

"Bald Eagle Attacks Two Boys." *New York Times,* May 9, 1895.

Barber, Elizabeth Wayland, and Paul T. Barber. *When They Severed Earth from Sky.* Princeton: Princeton University Press, 2004, pp. 110–11, 242.

Barber, J. W., and P. T. Barber. "Why the Flood Is Universal but Only Germanic Dragons Have Halitosis." *Proceedings of the Fifteenth Annual UCLA Indo-European Conference*, 2003.

Barber, Paul. *Vampires, Burial, and Death.* New Haven: Yale University Press, 1988, pp. 167–79, 190–98.

Bargu, Sibel, et al. "Mystery Behind Hitchcock's Birds." *Nature Geoscience* 5 (2003).

Barnes, Ian, et al. "Ancient Urbanization Predicts Genetic Resistance to Tuberculosis." *Evolution* 65, no. 3 (March 2011): 842–48.

Barnett, Ross, et al. "Phylogeography of Lions (*Panthera leo* ssp.) Reveals Three Distinct Taxa and a Late Pleistocene Reduction in Genetic Diversity." *Molecular Ecology* 18 (2009).

Bates, Roy. *Chinese Dragons.* New York: Oxford University Press, 2002.

Beckwith, Martha. *Hawaiian Mythology.* Honolulu: University of Hawaii Press, 1986.

Benchley, Peter. *Jaws.* Garden City, NY: Doubleday, 1974.

Beowulf. Translated by Francis B. Gummere. St. Petersburg, FL: Red and Black Publishers, 2007.

Billi, Andrea, et al. "On the Cause of the 1908 Messina Tsunami, Southern Italy." *Geophysical Research Letters* 35 (2008).

Bird, C. "The Sea-Serpent Explained." *Nature* 18 (1878): 519.

Bondeson, Jan. *Buried Alive.* New York: W. W. Norton, 2001, pp. 258–61.

Book of a Thousand Nights and a Night, The. Translated by Richard Burton. New York: Penguin, 2008.

Breasted, James Henry, ed. Breasted Ancient History Series. (Chicago: Denoyer-Geppert, 1931), Map. "Ancient Greece, Greek and Phoenician Colonies and Commerce." Map. 1931.

British Admiralty. "Mediterranean Sea, Italy, South Coast, Stretto di Messina." Map. 1945.

———. Pilot Guide—Mediterranean Sea. 2005, p. 422.

Buitron-Oliver, Diana, et al. *The "Odyssey" and Ancient Art.* Annandale-on-Hudson, NY: Bard College, 1992.

Bulfinch, Thomas. *Bulfinch's Mythology.* New York: Modern Library, 2004.

Burney, David A., et al. "A Chronology for Late Prehistoric Madagascar." *Journal of Human Evolution* 47 (July–August 2004): 25–63.

Callimachus. *Hymns and Epigrams.* Translated by A. W. Mair and G. R. Mair. Cambridge, MA: Harvard University Press, 1921.

Cameron, Ed. *The Psychopathology of the Gothic Romance.* Jefferson, NC: McFarland, 2010.

Chahal, Harvinder S., et al. "AIP Mutation in Pituitary Adenomas in the 18th Century and Today." *New England Journal of Medicine* 364 (January 6, 2011): 43–50.

Chamberlain, Ted. "Hogzilla Is No Hogwash." *National Geographic News,* March 22, 2005.

Chikhachev, P. A. "Carte Géologique de l'Asie Mineure." 1885.

Chippaux, J.-P. "Snake Bites: Appraisal of the Global Situation." *Bulletin of the World Health Organization* 76 (1998): 515–24.

Clarke, M. R., and N. Macleod. "Cephalopod Remains from Sperm Whales Caught off Iceland." *Journal of the Marine Biological Association of the United Kingdom* 56 (1976): 733–49.

Collins, Paul. "The Real Vampire Hunters." *New Scientist* 209 (January 2011): 40–43.

Cowen, Richard. *History of Life.* New York: McGraw-Hill, 1976.

Crichton, Michael. *The Andromeda Strain.* New York: Knopf, 1969.

———. *Jurassic Park.* New York: Knopf, 1990.

Dahlitz, M., and J. D. Parkes. "Sleep Paralysis." *Lancet* 341 (February 1993): 406–7.

Davis, E. Wade. "The Ethnobiology of the Haitian Zombi." *Journal of Ethnopharmacology* 9 (1983): 85–104.

———. *Passage of Darkness.* Chapel Hill: University of North Carolina Press, 1988, pp. 156–59.

"Deaths by Cause, Sex and Mortality Stratum." World Health Report, 2004.

Diedrich, Cajus G. "Upper Pleistocene *Panthera leo spelaea* Remains from the Bilstein Caves and Contribution to the Steppe Lion Taphonomy, Palaeobiology and Sexual Dimorphism." *Annales de Paléontologie* 95 (July–September 2009): 117–38.

Doyle, Arthur Conan. *Lost World.* New York: Penguin 2002.

Eason, Cassandra. *Fabulous Creatures, Mythical Monsters, and Animal Power Symbols.* Westport, CT: Greenwood, 2008, pp. 19–30, 35–37, 59, 62–63, 110–12.

Eastwick, Paul W., and Eli J. Finkel. "Sex Differences in Mate Preferences Revisited." *Journal of Personality and Social Psychology* 94 (2008): 245–64.

Ellis, Richard. *Monsters of the Sea.* New York: Knopf, 1994, pp. 260–63, 372–76.

Epic of Gilgamesh, The. Translated by N. K. Sandars. New York: Penguin, 1977.

Euripides. *Orestes.* Translated by E. P. Coleridge. New York: Random House, 1938.

———. *Medea.* Translated by John Davie. New York: Penguin, 1996.

Evans, W. E. D. *The Chemistry of Death.* Springfield, IL: Charles C. Thomas, 1963.

Fast, Nathanael J., et al. "The Destructive Nature of Power Without Status." *Journal of Experimental Social Psychology,* 2011.

Fernicola, Richard G. *Twelve Days of Terror: A Definitive Investigation of the 1916 New Jersey Shark Attacks.* Guilford, CT: Lyons Press, 2001.

Finkel, Eli J., and Paul W. Eastwick. "Speed-Dating." *Current Directions in Psychological Science* 17, no. 3 (2008): 193–97.

Finkel, Eli J., Paul W. Eastwick, and Jacob Matthews. "Speed-Dating as an Invaluable Tool for Studying Romantic Attraction." *Personal Relationships* 14 (2007): 149–66.

Flöistrup, Bertil. *The World of Dragons.* Hong Kong: Joint Publishing, 2000.

Florida Museum of Natural History. Shark Attack Database. www.flmnh.ufl.edu/fish/sharks/isaf/isaf.htm.

Freud, Sigmund. *Totem and Taboo.* Translated by A. A. Brill. Mineola, NY: Dover, 1983, pp. 1–5.

Fry, Bryan G., et al. "Early Evolution of the Venom System in Lizards and Snakes." *Nature* 439 (February 2, 2006): 584–88.

Ganas, A., and T. Parsons. "Three-Dimensional Model of Hellenic Arc Deformation and Origin of the Cretan Uplift." *Journal of Geophysical Research* 114 (2009).

Garamszegi, László, Marcel Eens, and János Török. "Birds Reveal Their Personality When Singing." *PLoS ONE* 3, no. 7 (2008).

Garrison, Tom. *Oceanography: An Invitation to Marine Science.* 6th ed. Belmont, CA: Brooks/Cole, 2007.

Geoffrey of Monmouth. *History of the Kings of Britain.* Translated by Sebastian Evans. New York: Dutton, 1958.

Gizi Map. Silk Road Countries.

Gómez-Alonso, Juan. "Rabies: A Possible Explanation for the Vampire Legend." *Neurology* 51, no. 3 (September 1, 1998): 856–59.

Gould, Charles. *Mythical Monsters.* Detroit: Singing Tree Press, 1969.

Gunduz, A., et al. "Wild Boar Attacks." *Wilderness Environmental Medicine* 18, no. 2 (Summer 2007): 117–19.

Hadly, Elizabeth A., and Anthony B. Barnosky. "Vertebrate Fossils and the Future of Conservation Biology." *Conservation Paleobiology* 15 (October 2009).

Hall, Howard. "Mugged by a Squid." *Ocean Realm* (1991): 6–8.

Hay, John. *Ancient China*. London: Bodley Head, 1973.

Hayden, Thomas. "How to Hatch a Dinosaur." *Wired*, September 26, 2011.

Headland, Thomas N., and Harry W. Greene. "Hunter-Gatherers and Other Primates as Prey, Predators, and Competitors of Snakes." *Proceedings of the National Academy of Sciences* 108 (December 2011).

Hemenover, Scott H., and Ulrich Schimmack. "That's Disgusting! . . . But Very Amusing: Mixed Feelings of Amusement and Disgust." *Cognition and Emotion* 21, no. 5 (2007): 1102–13.

Herborn, K. A., et al. "Oxidative Profile Varies with Personality in European Greenfinches." *Journal of Experimental Biology* 214 (2011): 1732–39.

Herodotus. *The Histories*. Translated by George Rawlinson. Rutland, VT: Tuttle, 1992.

Hesiod. *The Homeric Hymns and Homerica*. Translated by Hugh G. Evelyn-White. Cambridge, MA: Harvard University Press, 1982.

Hill, Kim, and A. Magdalena Hurtado. *Aché Life History*. New York: Aldine de Gruyter, 1996, pp. 1–4, 162–63.

Hoffman, Paul F., and Daniel P. Schrag. "Snowball Earth." *Scientific American*, January 2000, pp. 68–75.

Hoffman, Paul F., et al. "A Neoproterozoic Snowball Earth." *Science* 281 (August 28, 1998): 1342–46.

Homer. *The Iliad of Homer*. Translated by Richmond Lattimore. Chicago: University of Chicago Press, 2011.

———. *The Odyssey of Homer*. Translated by Richmond Lattimore. New York: Harper, 1967.

———. *The Odyssey*. Translated by Samuel Butler. Wildside Press, 2007.

Hoult, Janet. *Dragons*. Glastonbury, UK: Gothic Image Publications, 1987, pp. 52–55, 86–89.

Huang H. *Commentary on the Lun Heng and Wang Ch'ung*. Beijing: Zhonghua Book Company, 1990.

Hyde, William T., et al. "Neoproterozoic 'Snowball Earth' Simulations with a Coupled Climate/Ice-Sheet Model." *Nature* 405 (May 25, 2000): 403–4.

Idel, Moshe. *Golem: Jewish Magical and Mystical Traditions on the Artificial Anthropoid*. Albany, NY: State University of New York Press, 1990.

Italian Government. Mediterranean Sea, Italy, the Faro or Strait of Messina. Map. 1881.

Jungebluth, Philipp, et al. "Tracheobronchial Transplantation with a Stem-Cell-Seeded Bioartificial Nanocomposite." *Lancet* 378 (December 10, 2011): 1997–2004.

Kaplan, Colin, ed. *Rabies: The Facts.* Oxford: Oxford University Press, 1977, pp. 38–41.

Kaplan, Jonathan E., James W. Larrick, and James A. Yost. "Snake Bite Among the Waorani Indians of Eastern Ecuador." *Transactions of the Royal Society of Tropical Medicine and Hygiene* 72, no. 5 (1978).

Kaplan, Matt. "Let the River Run." *New Scientist* 175 (September 28, 2002).

———. "Why Some Animals Are Shy of Habitat Corridors." *Nature,* June 12, 2007.

———. "Moose Use Roads as a Defence Against Bears." *Nature,* October 10, 2007.

———. "Snakes' Venom Chemistry Varies with Age and Location." *Nature,* July 16, 2008.

———. "You Have 4 Minutes to Choose Your Perfect Mate." *Nature* 451 (2008): 760–62.

———. "Zoos Help Track Spread of Pandemics." *Nature,* May 4, 2009.

———. "Unnatural Selection." *Economist,* May 21, 2009.

———. " 'Sea Monster' Bones Reveal Ancient Shark Feeding Frenzy." *National Geographic,* September 28, 2009.

———. "The Snake That Swallowed Dinosaurs." *Nature,* March 2, 2010.

———. "Blog Mining." *Economist,* March 11, 2010.

———. "From Gollum to Avatar." *Economist,* June 10, 2010.

———. "Nods and Winks." *Economist,* June 10, 2010.

———. "The Men Who Stare at Men." *New Scientist* 208 (November 6, 2010): 38–39.

———. "When Snake Fangs Moved out of the Groove." *Nature,* November 17, 2010.

———. " 'Zombie' Ants Found with New Mind-Control Fungi." *National Geographic,* February 9, 2011.

———. "Early Europeans Unwarmed by Fire." *Nature,* March 14, 2011.

———. "Painted Out." *Economist,* March 17, 2011.

———. "Zombie Power: Harnessing Parasite Mind Control." *New Scientist,* August 30, 2011.

———. "Giving Drones a Thumbs Up." *Economist,* March 24, 2012.

———. "Time to Be Honest." *Economist,* March 31, 2012.

Keizer, Kees, Siegwart Lindenberg, and Linda Steg. "The Spreading of Disorder." *Science* 322 (December 2008): 1681–85.

Kelly, Gavin. "Ammianus and the Great Tsunami." *Journal of Roman Studies* 94 (2004): 141–67.

King, Glenn E. "The Attentional Basis for Primate Responses to Snakes." Paper presented at the Annual Meeting of the American Society of Primatologists, 1997.

King, Stephen. *'Salem's Lot.* Garden City, NY: Doubleday, 1975.

———. *The Shining.* Garden City, NY: Doubleday, 1977.

———. *Pet Sematary.* Garden City, NY: Doubleday, 1983.

Kolata, Gina. "In a Giant's Story." *New York Times,* January 5, 2011.

Kryger, M. H. *International Classification of Sleep Disorders.* Philadelphia: W. B. Saunders, 2000, pp. 166–69.

Kryger, Meir H., Thomas Roth, and William C. Dement. *Principles and Practice of Sleep Medicine.* 5th ed. St. Louis, MO: Elsevier/Saunders, 2011.

Lafferty, Kevin D. "Can the Common Brain Parasite, *Toxoplasma gondii,* Influence Human Culture?" *Proceedings of the Royal Society* 273 (November 2006): 2749–55.

"Legacy of Doctor Moreau, The." *Nature,* July 22, 2011.

Lehner, Ernst. *A Fantastic Bestiary: Beasts and Monsters in Myth and Folklore.* New York: Tudor Publishing, 1969, pp. 21–38, 44–59, 65–69, 73, 110–11.

Lurker, Manfred. *The Routledge Dictionary of Gods, Goddesses, Devils and Demons.* New York: Routledge, 2004.

Marean, C. W., et al. "Early Human Use of Marine Resources and Pigment in South Africa During the Middle Pleistocene." *Nature* 449 (October 18, 2007): 905–8.

Margalida, A. "Bearded Vultures (*Gypaetus barbastus*) Prefer Fatty Bones." *Behavioral Ecology and Sociobiology* 63, no. 2 (2008): 187–93.

Martin, Brian P. *World Birds.* Enfield, UK: Guinness Books, 1987, pp. 127–28.

Mayor, Adrienne. *The First Fossil Hunters.* Princeton: Princeton University Press, 2000, pp. 34–36, 38–53, 121–26.

McGraw, W. S., C. Cooke, and S. Shultz. "Primate Remains from African Crowned Eagle (*Stephanoaetus coronatus*) Nests in Ivory Coast's Tai Forest." *American Journal of Physical Anthropology* 131, no. 2 (October 2006): 151–65.

McKay, David S., et al. "Search for Past Life on Mars: Possible Relic Biogenic Activity in Martian Meteorite ALH84001." *Science* 273 (August 16, 1996): 924–30.

Meijaard, Erik, and Douglas Sheil. "A Modest Proposal for Wealthy Countries to Reforest Their Land for the Common Good." *Biotropica* 43 (September 2011): 524–28.

Melin, Amanda D., et al. "Effects of Colour Vision Phenotype on Insect Capture by a Free-ranging Population of White-faced Capuchins." *Animal Behaviour* 73 (2007): 205–14.

Mineka, Susan, Richard Keir, and Veda Price. "Fear of Snakes in Wild- and Laboratory-Reared Rhesus Monkeys (*Macaca mulatta*)." *Learning & Behavior* 8, no. 4 (1980): 653–63.

Mitchell, J. S., A. B. Heckert, and H. D. Sues. "Grooves to Tubes: Evolution of the Venom Delivery System in a Late Triassic 'Reptile.'" *Naturwissenschaften* 97, no. 12 (December 2010): 1117–21.

Morales, Judith, Roxana Torres, and Alberto Velando. "Safe Betting: Males Help Dull Females Only When They Raise High-quality Offspring." *Behavioral Ecology and Sociobiology* 66 (2011): 135–43.

Morency, Louis-Philippe, Iwan de Kok, and Jonathan Gratch. "A Probabilistic Multimodal Approach for Predicting Listener Backchannels." *Journal of Alternative Agent and Multiagent Systems* 20, no. 1 (2010).

Morency, Louis-Philippe, Rada Mihalcea, and Payal Doshi. "Towards Multimodal Sentiment Analysis." *Proceedings of the International Conference on Multimodal Interfaces*, 2011.

Murgatroyd, Paul. *Mythical Monsters in Classical Literature.* London: Duckworth, 2007, pp. 44–48.

Ness, Robert C. "The Old Hag Phenomenon as Sleep Paralysis." *Culture, Medicine, and Psychiatry* 2, no. 1 (March 1978): 15–39.

Nihon, N. I. "Activities of Toxic Component of Venom in Snake Bite." *Japanese Society of Tropical Medicine* 12, no. 2 (1984).

O'Brien, Shaun. "Skeletons in the Campanile's Closet." *Daily Californian*, January 24, 2001.

Öhman, Arne, and Susan Mineka. "Fear, Phobias, and Preparedness: Toward an Evolved Module of Fear and Fear Learning." *Psychological Review* 108, no. 3 (2001): 483–522.

———. "The Malicious Serpent: Snakes as a Prototypical Stimulus for an Evolved Module of Fear." *Current Directions in Psychological Science* 12 (2003): 2–9.

Öhman, A., A. Flykt, and F. Esteves. "Emotion Drives Attention: Detecting the Snake in the Grass." *Journal of Experimental Psychology: General* 130, no. 3 (September 2001): 466–78.

Ovid. *Metamorphoses.* Translated by A. D. Melville. New York: Oxford University Press, 1998.

Owen, James. "Loch Ness Sea Monster Fossil a Hoax, Say Scientists." *National Geographic News,* July 29, 2003.

Paolini, Christopher. *Eragon.* New York: Knopf, 2003.

Papadopoulos, John K., and Deborah Ruscillo. "A *Ketos* in Early Athens: An Archaeology of Whales and Sea Monsters in the Greek World." *American Journal of Archaeology* 106 (2002): 187–227.

Parry, Vivienne. "How an Irish Giant and an 18th-Century Surgeon Could Help People with Growth Disorders." *Guardian,* January 10, 2011.

Perrett, David I., et al. "Symmetry and Human Facial Attractiveness." *Evolution and Human Behavior* 20, no. 5 (September 1999): 295–307.

Petroleum Economist. Energy Map of the Middle East and Caspian. 2000.

Petronius. *The Satyricon.* Translated by Patrick Gerard Walsh. New York: Oxford University Press, 2009.

Phemister, Dallas B., and Huberta Livingstone. "Primary Shock." *Annals of Surgery* 100, no. 4 (October 1934): 714–27.

Pindar. *The Odes.* Translated by G . S. Conway. London: Orion Publishing, 1998.

Pindar. *The Odes of Pindar.* Translated by Sir John Sandys. Cambridge, MA: Harvard University Press, 1937.

Plato. *Plato in Twelve Volumes* 3. Translated by W. R. M. Lamb. Cambridge, MA: Harvard University Press, 1967.

Polo, Marco. *The Travels of Marco Polo.* 2 vols. Translated by Henry Yule. New York: Dover, 1993.

Rhodes, Gillian, and Leslie A. Zebrowitz, eds. *Facial Attractiveness: Evolutionary, Cognitive, and Social Perspectives.* Westport, CT: Ablex Publishing, 2001.

Roebroeks, Wil, and Paola Villa. "On the Earliest Evidence for Habitual Use of Fire in Europe." *Proceedings of the National Academy of Sciences* 108, no. 13 (March 14, 2011): 5209–14.

Rosa, Rui, and Brad A. Seibel. "Slow Pace of Life of the Antarctic Colossal Squid." *Journal of the Marine Biological Association of the United Kingdom* 90 (2010): 1375–78.

Rosenberg, Yudi. *The Golem and the Wondrous Deeds of the Maharal of Prague.* New Haven: Yale University Press, 2007.

Ross, Andrew. *Amber: The Natural Time Capsule.* London: Natural History Museum, 1998, pp. 32–33.

Ross, Andrew, and Jeremy Austin. "The Search for DNA in Amber." London: Natural History Museum, 1998. www.nhm.ac.uk/resources-rx/files/12feat_dna_in_amber-3009.pdf.

Rowling, J. K. *Harry Potter and the Philosopher's Stone.* New York: Scholastic, 1998.

———. *Harry Potter and the Chamber of Secrets.* New York: Scholastic, 1999.

———. *Harry Potter and the Prisoner of Azkaban.* New York: Scholastic, 1999.

———. *Harry Potter and the Goblet of Fire.* New York: Scholastic, 2000.

———. *Harry Potter and the Deathly Hallows.* New York: Scholastic, 2007.

Rozin, Paul, and Deborah Schiller. "The Nature and Acquisition of a Preference for Chili Pepper by Humans." *Motivation and Emotion* 4, no. 1 (1980): 77–101.

Rudwick, Martin J. S. *The Meaning of Fossils.* Chicago: University of Chicago Press, 1985.

Sagan, Carl. *The Dragons of Eden.* New York: Random House, 1977, pp. 131–38, 142.

Sanders, Tao Tao Liu. 1980. *Dragons, Gods and Spirits from Chinese Mythology.* New York: Schocken, 1983, pp. 48–49.

Scheffers, Anja, et al. "Late Holocene Tsunami Traces on the Western and Southern Coastlines of the Peloponnesus (Greece)." *Earth and Planetary Science Letters* 269 (2008): 271–79.

Schulp, Anne S., Mohammed Al-Wosabi, and Nancy J. Stevens. "First Dinosaur Tracks from the Arabian Peninsula." *PLoS ONE* 3, no. 5 (2008).

Scofield, R. P., and K. W. S. Ashwell. "Rapid Somatic Expansion Causes the Brain to Lag Behind: The Case of the Brain and Behavior of New Zealand's Haast's Eagle (*Harpagornis moorei*)." *Journal of Vertebrate Palaeontology* 29, no. 3 (2009).

Shalvi, Shaul, Ori Eldar, and Yoella Bereby-Meyer. "Honesty Requires Time (and Lack of Justification)." *Psychological Science,* 2012.

Shaw, B., et al. "Eastern Mediterranean Tectonics and Tsunami Hazard Inferred from the AD 365 Earthquake." *Nature Geoscience* 1 (2008): 268–76.

Shelley, Mary. *Frankenstein.* New York: Penguin, 2003.

Shirihai, Hadoram. *Whales, Dolphins and Seals.* London: A. & C. Black, 2006.

Shuker, Karl. *Dragons: A Natural History.* New York: Simon & Schuster, 1995.

Smyth, W. H. "Plan of the Faro, or Strait of Messina." Map. British Library, 1823.

Solounias, Nikos, and Adrienne Mayor. "Ancient References to the Fossils from the Land of Pythagoras." *Earth Sciences History* 23, no. 2 (2004): 283–96.

Song, Yale, et al. "Continuous Body and Hand Gesture Recognition for Natural Human-Computer Interaction." *ACM Transactions on Interactive Intelligent Systems* 2, no. 5 (March 1012).

Standard Oil Co. "The Middle East Oil Industry." Map. 1950.

Stewart, Charles. "Erotic Dreams and Nightmares from Antiquity to the Present." *Journal of the Royal Anthropological Institute* 8 (2002): 279–309.

Stewart, Kathlyn M. "Fishing Sites of North and East Africa in the Late Pleistocene." *British Archaeological Reports* 34 (1989).

Stoker, Bram. *Dracula.* New York: Penguin, 2003.

Stothers, R. B. "Ancient Scientific Basis of the 'Great Serpent' from Historical Evidence." *Isis* 95 (2004): 220–38.

Strabo. *Geography.* 8 vols. Translated by Horace Leonard Jones. Cambridge, MA: Harvard University Press, 1917–1932.

Swaroop, S., and B. Grab. "The Global Burden of Rabies and Envenomings." *Bulletin of the World Health Organization* 10 (1954).

Théodoridès, Jean. *Histoire de la Rage: Cave Canem.* Paris: Masson, 1986, pp. 22–23, 100–7.

Thorp, Robert L. *China in the Early Bronze Age: Shang Civilization.* Philadelphia: University of Pennsylvania Press, 2006.

Tolkien, J. R. R. *The Hobbit.* Boston: Houghton Mifflin, 2001.

———. *The Lord of the Rings.* Boston: Houghton Mifflin, 2004.

Trigg, Elwood B. *Gypsy Demons and Divinities: The Magic and Religion of the Gypsies.* Secaucus, NJ: Citadel Press, 1973, pp. 148–55.

Verne, Jules. *20,000 Leagues Under the Sea.* New York: Tom Doherty Associates, 1995.

———. *Journey to the Centre of the Earth.* New York: Oxford University Press, 1998.

Vukanovi, Tatomir P. "The Vampire." *Journal of the Gypsy Lore Society* 36 (1957): 128–31.

Vyasa. *Mahabharata.* Retold by William Buck. Berkeley: University of California Press, 1973.

Webster, J. P., and G. A. McConkey. "*Toxoplasma gondii*–altered Host Behaviour: Clues as to Mechanism of Action." *Folia Parasitologica* 57, no. 2 (June 2010): 95–104.

Wells, H. G. *The War of the Worlds.* New York: Penguin, 2005.

———. *The Island of Doctor Moreau.* New York: Penguin, 2010.

William of Newburgh. *The History of English Affairs.* Translated by P. G. Walsh and M. J. Kennedy. Warminster: Aris & Phillips, 1988.

Winkler, Gershon. *The Golem of Prague.* New York: Judaica Press, 1980.

Witton, M. P., and D. Naish. "A Reappraisal of Azhdarchid Pterosaur Functional Morphology and Paleoecology." *PLoS ONE* 3, no. 5 (2008).

Zaidel, D. W., S. M. Aarde, and K. Baig. "Appearance of Beauty, Health, and Symmetry in Human Faces." *Brain and Cognition* 57, no. 3 (2005): 261–63.

Zhong, Chen-Bo, Vanessa K. Bohns, and Francesca Gino. "A Good Lamp Is the Best Police: Darkness Increases Dishonesty and Self-Interested Behavior." *Psychological Science* 21 (2010): 311–14.

Index

Page numbers in *italics* refer to illustrations.